Creative
Machine
Art

Creative
Machine Art

Sharee Dawn Roberts

American Quilter's Society
P.O. Box 3290, Paducah, KY 42002-3290

Library of Congress Cataloging-in-Publication Data

Roberts, Sharee Dawn.
    Creative machine art/Sharee Dawn Roberts.
    p. cm.
    ISBN 0-89145-986-3: $24.95
    1. Machine sewing. 2. Machine appliqué. 3. Embroidery, Machine.
I. Title
TT713.R63   1992   92-9409
746.44–dc20   CIP

Additional copies of this book may be ordered from:

American Quilter's Society
P.O. Box 3290
Paducah, KY 42002-3290

@$24.95. Add $1.00 for postage & handling.

DEDICATION

To my Parents
For encouraging me to dream,
and for teaching me the value of working hard
to attain my dreams.

ACKNOWLEDGMENTS

I wish to express my heartfelt thanks to the many people who helped make this book a reality. Thank you, Bill and Meredith Schroeder, for having confidence in me to write a fine machine art book, and Victoria Faoro for her very encouraging support. I wish to acknowledge Beth Summers, Elaine Wilson and Kay Smith for contributing their creative talents in the production of the visuals and the manuscript. I am grateful to photographer Glenn Hall for supplying his expertise, and to Joanna Pace for being such a fine model for my clothing. A very special thanks to photographer Elizabeth Courtney, who seemed to sense instinctively just what I wanted to capture on film. I would like to give special recognition and appreciation to the contributing artists for sharing their talents: Rosemary Ponte, Judy Simmons, Jeanie Sexton, Paulette Thompson, Beverly Burton, Linda Stallion, Marilyn Boysen, Lorelle LaBarge and Karen Ballash. I am grateful to Pfaff™, for developing a sewing machine that responds to the special demands of the fiber artist. I feel indebted to Linda and Keith English, who have shown their faith in me from the very beginning and continue to enthusiastically sponsor my classes.

I shall always be grateful to my friend Kathy Hicks for introducing me to the world of sewing and giving me the confidence to learn.

A very loving thanks to my sisters, who think everything I create is wonderful. I thank my children for putting up with my eccentricities, and my husband for being my best critic and helpmate.

Most of all, I would like to thank all of my students for their support, especially my local students at English Sewing Center, for without their encouragement and inspiration, this book would never have been written.

Note: All work shown is by the author unless otherwise noted.

Introduction

When I announce that I am a "sewing machine artist" and someone turns up his/her nose, I cannot feel indignant. Instead I am amused, because I can relate so personally. I was the world's worst "snob" about machine work, because I considered handwork to be the only acceptable stitchery. I had been exposed to only poor and mediocre samples of machine appliqué. I was a novice on the sewing machine myself, using it only for sewing together pillow tops or quick patterns. But I would spend hours every day lavishing intense attention on hand piecing, hand quilting, hand appliqué, and hand embroidery.

I probably would never have considered exploring the merits of the sewing machine without the prompting of a good friend of mine. She had taken a course in machine embroidery from a local shop and was so enthralled with what she had learned that she literally signed me up for the next set of lessons. All the while I was thinking to myself, "tacky, tacky," but agreed to attend the classes because some other friends were going as well. It was an excuse to get out of the house with the "girls," a nice break from my two active toddlers.

Lives sometimes change overnight. When I walked into the classroom and saw the

instructor's beautifully crafted samples, I was astounded! I suspiciously thought to myself that she must have completed part of the work by hand. I was not convinced she hadn't until she sat down to demonstrate. As she sat behind her sewing machine, effortlessly guiding the hoop under the needle, she created a beautifully shaded flower right before my eyes. At that very moment I was consumed with a burning desire to create beautiful work with my sewing machine, a desire that will remain with me for life!

The machine embroidery class was a failure as far as my learning the skill because I had just purchased a new sewing machine and barely knew how to thread it. I was mortified by the thought of making any tension adjustments and simply could not remember to lower my presser bar while embroidering. No matter, the seed of determination was planted. After graciously "flunking" the three short lessons, I spent the next couple of weeks teaching myself to operate my sewing machine, and then gradually experimented with appliqué and embroidery techniques. I have not stopped experimenting since taking that class eight years ago because the possibilities for machine art are truly infinite!

Let us consider the sewing machine for a few moments. For those who are convinced it is an inferior substitute for fine handwork, let me ask this question: how fine would handwork be without the sharp steel needle used to penetrate the fabric, allowing the thread to pass through to create the highly

BELOW: A sample swatch of experimental stitches. Working upside down with a very tight bobbin tension and a very loose top tension will produce a sampler such as this one. The rayon embroidery thread from the top is pulled through to the tightened bobbin that is filled with regular thread, forming very interesting whirls and spokes.

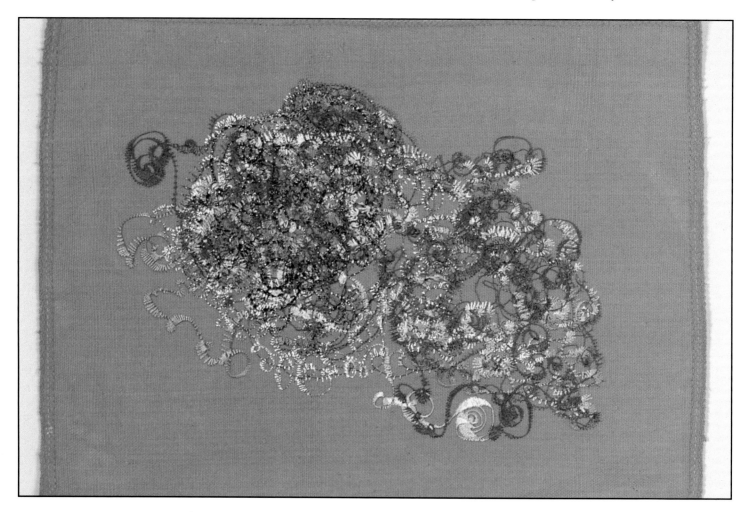

praised embroidery stitches? Is not that needle, too, a "tool" invented by man and manufactured by machines in order to facilitate stitching the designs the artist has envisioned? Well, the sewing machine is simply another tool. Indeed, it is much larger and more complex than a hand needle, but it is certainly not a tool that is simply put on "automatic pilot." Without skillful human hands and a creative mind at work, this tool would sit as useless as would a little steel hand needle with no skillful hands to guide it.

Hopefully, machine stitchery will someday be as highly regarded as fine hand stitchery.

One thing is very clear: machine stitchery has its own style and personality and should not be used to reproduce or imitate hand stitchery. The two techniques are as different as watercolor and oil painting are to each other, and the end product will be as different as well.

Individual style evolves under the hands of the artist. Though this book includes instructions for developing technical skill, the artist should not try to copy the illustrations exactly. Think of the sewing machine as a substitute for the artist's paintbrush or charcoal stick; machine artists' styles of directing thread will differ as dramatically as the

brush strokes of two different painters. This is to be expected, not avoided, and is one reason the world of sewing is such an exciting arena for the artist.

It's my hope that this book not only teaches the mechanics of appliqué and embroidery techniques, but also inspires the artist to develop a personal style. Every individual possesses an innate sense of design, and every person's work will be an individualistic statement. The sewing machine is nothing more than a tool to fabricate this creative urge, a tool that I sincerely hope will give much enjoyment and at the same time fulfill the artist's need for expression.

1. The Sewing Machine

Had someone told me eight years ago that someday I'd be writing a book acclaiming the merits of the sewing machine, I would have answered, "Yes, and someday I will become the first woman president of the United States!" The idea would have seemed especially preposterous because I did not learn sewing at an early age. In fact, I had a mental block about sewing in general because of the bad experience I had in a mandatory home economics course I took in seventh grade. I had no interest in sewing at that time, which

LEFT: The sewing machine becomes an extension of the machine artist's hands.

was reflected in my projects. I seemed to run into one huge obstacle after another. The whole world of sewing seemed so complex and alien that I just didn't have the *motivation* to learn. After that class I vowed to never go near a sewing machine again; it was all too difficult! Besides that, all the good seamstresses I knew had learned to sew at their mothers' knees at age five. It seemed far too late for me!

I was not re-introduced to sewing again until I was a junior at San Diego State University where I was an art student majoring in drawing and painting. I had taken an elective course in textile design because

I wanted to learn to weave. That first semester of textile design opened up a whole new world for me. I not only explored weaving, but also block printing, silkscreening, batik and tie-dyeing. I was fascinated by the new dimension textiles offered and painting did not; that is, of course, *texture*. I kept repeating more courses in textile design and eventually switched my emphasis to textiles. I was having the time of my life printing yards and yards of the most intriguing designs. Not once did it occur to me to consider what to do with all this hand-printed yardage; I was creating for the pure joy of designing! Then one semester my professor announced that our final project would be the designing and printing of a garment.

A garment? You mean one put together by *seams*? Seams stitched on a *sewing machine*? No one but a professed non-sewer could know the panic I experienced. I felt totally helpless and incapable, and at that point realized what a ridiculous position I was in, as a non-sewing textile design major!

Fortunately for me, I had made friends with a girl in my class who could look at a garment, go home and stitch it up, improving it in the process. She was the most expert seamstress I knew; I marveled at her expertise. Her "home-sewn" garments were the first I had ever seen that surpassed the finest designer wear found in fashionable boutiques. She noted my despair, took me under her wing and encouraged me to try. She did not teach me *how* to sew (with full-time school and young children at home, there simply wasn't time!). But I'll forever be grateful to her for making me feel as though I could do it. I can still hear her prompting me, "You can do it; it is so easy!" I secretly smile when I now find myself repeating that same phrase to my students. My experience has confirmed my belief that all that is needed for success is the will to try and determination!

I finally had the motivation and the confidence to sew; all I needed was simply to learn how. Anyone can teach him or herself how to sew with the many excellent books available. I also relied on the instructions and illustrations of the patterns I used. The "know-how," I found, was not my biggest obstacle. It was the sewing machine. I still harbored ill feelings toward this mechanical fabric-eater. My first mistake was purchasing a used sewing machine. Knowing absolutely nothing about sewing, I would have been far wiser to have gone to a reputable dealer and learned how to operate the machine with a professional. So my trial-and-error was far more error than trial, and I quickly became very frustrated. I didn't give up because I didn't have a choice; I had to learn. I stuck with straight seams for two years because that was about the extent of my machine's capability. I sewed together "Sew-Easy" patterns and quilt blocks.

Finally I traded my "doubly-used" sewing machine for a good, new one sold by a reliable dealer. I faithfully attended the free lessons offered to new owners, and I marveled at the additional functions my new machine could perform. It was at this time that I sauntered unaware into the machine embroidery class. My motivation to sew creatively became even stronger than my desire to learn the fundamentals of dressmaking. I wanted to create those gorgeous appliquéd and embroidered designs with an almost fierce determination.

I spent literally days producing nothing but scraps of experimental stitches. I've always been a "process" person and never had a compelling need to produce "things." To learn a new skill was enough to satisfy my creative drive. I feel students would benefit more if they were as willing to "play and learn" as they are to produce. When learning machine skills it is very important to spend many hours practicing and experimenting. Just remember, as with all skills, the hand and mind must learn to function as a unit, and this simply takes practice. But, also like all other skills, once mastered, it becomes a very natural, almost unconscious process. Perhaps that is why I repeat, "It is so easy" to my students. It does become so easy that you will believe the sewing machine is an extension of your hands!

No one can talk objectively about sewing machines. Those who love to sew quite naturally love their sewing machines. If you are reading this with the intention of learning the machine arts, but think of your sewing machine as a "monster," then my advice is to *sell* it. You must have a loving relationship with your sewing machine, and some

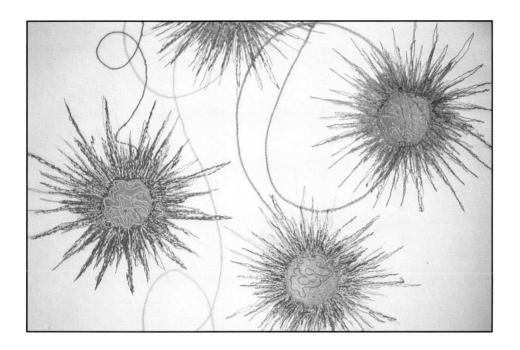

machines and some people are just not compatible. A sewing machine is just like a car or a house; certain features or styles are enticing to some people but not to others. I have worked with enough students and their different machines to realize what a personal decision it becomes. That is why there is no "best" sewing machine on the market! What is important is to shop wisely, and test drive many different machines before making a final decision. Make a written list of features that are desirable and those that are mandatory. When shopping for a sewing machine, test those features.

Assuming the reader wants to utilize his/her sewing machine for the machine arts, these are the features I would recommend the sewing machine have:

- Zigzag with easy to manipulate, gradated stitch width settings.
- Feed-dog drop. (Cover plates are very awkward.)
- Easily adjustable top tension.
- Removable metal bobbin case, with adjustable tension screw.
- Horizontal thread pin. (Specialty threads will feed more smoothly.)
- "Needle down" button.
- Built-in needle threader.
- Twin needle capability.
- "Tie-off" button. (Automatically ties thread ends for appliqué and embroidery.)

Be sure to test your machine for the fine satin stitch used in appliqué. I find that this is the one performance that varies most from machine to machine. Some sewing machines simply will not produce a close satin stitch without stacking or feeding unevenly. Take different types of fabric with you and thoroughly test the machines. The machine should feed with absolutely no help, and create smooth, satiny stitches. Also check the tapering capabilities. You will want the stitch to be able to gradate from the widest width setting all the way down to the narrowest in a very smooth taper. Ideally, the width settings will be marked by numbers. This way you can exactly reproduce a particular width by associating it with a particular number.

Another consideration is the dealership you buy your sewing machine from. They should offer free lessons to help you learn all about your sewing machine. They should stand behind their sewing machines and be able to service them without shipping them off. They should be pleas-

ant and available for advice and offer occasional sewing seminars. Having a personal relationship with your sewing machine dealership can be very valuable as they can keep you informed about current trends, new techniques, attachments, sewing aids and products. Many dealers stock VCR tapes on machine embroidery, appliqué, cut-work, French hand sewing (by machine), etc. These excellent tapes contain a wealth of information, so take advantage of this service.

It is not necessary to purchase the most expensive sewing machine to create beautiful machine stitchery. But do evaluate what are the most important functions for you, and make sure you are going to be happy with the sewing machine you bring home!

Once you have purchased your sewing machine, do take advantage of the free lessons to learn its functions and maintenance. Read the owner's manual from front to back and over again. Learn how to attach and use all the different feet and any other attachments. Spend lots of time really learning how to operate your machine. Learn how to oil and clean your machine, for machine embroidery and appli-

LEFT: Art created by machine will have a distinctive style never meant to imitate hand stitchery.

ABOVE: Realistic images come alive beneath the sewing machine needle.

qué will accumulate lots of "fuzz balls" that will need to be kept clear of the bobbin area and underneath the throat plate. The more comfortable you are with your machine's operation, the more quickly you can relax and enjoy utilizing its very unique capabilities for machine art!

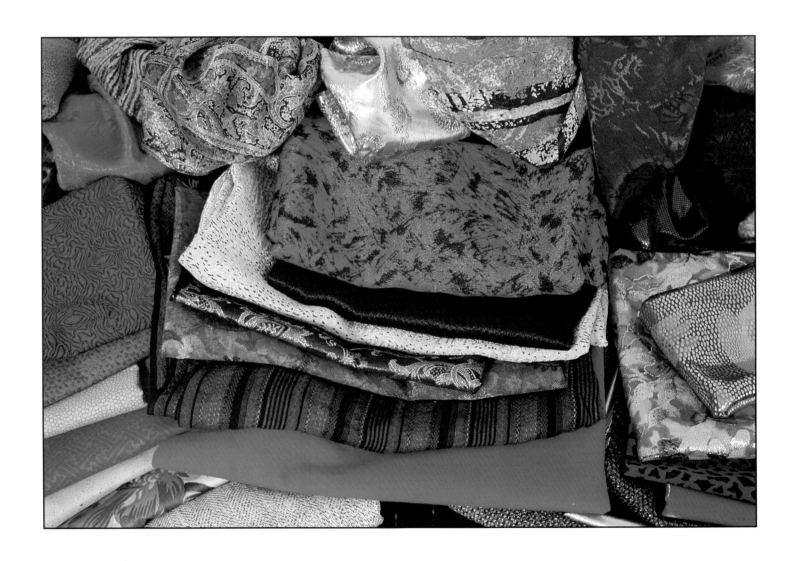

2. Fabric

An entire chapter is being devoted to this essential ingredient for machine art, to emphasize the important role fabric plays. Indeed, I am truly passionate about the subject and cannot refer to fabric simply as a "supply." I love fabric, all fabric, and there is not one in existence that I would rule out for machine art. I scout the drapery, bridal, upholstery, craft, quilting and dressmaking departments of fabric stores. But my unquenchable search does not end there! I head for the neglected clothes racks at yard sales, and feel my

LEFT: Fabric provides a wealth of exciting colors, designs and textures to inspire the machine artist.

blood rise in anticipation of finding treasures at our local Salvation Army store. I rummage through old trunks and attics, I collect and treasure ethnic and vintage clothing, and I linger over displays of antique and imported lace. But this is still not an inclusive list, for I consider anything paper or cloth-like a textile suitable for appliqué. I just as eagerly collect such objects as feathers, preserved flowers and weeds, rice papers, wallpaper, carpet remnants, plastics, leather and vinyl. I will not attempt to list all the possibilities; by now you have the general idea – do not be afraid to try anything! Of course one would not use paper or a similar textile

in an item to be washed or handled. Eventual use must be considered when you are selecting suitable appliqué materials.

The single element a fabric possesses that intrigues me most is its inherent texture. A painter relies on color, the musician on timing and scale, but the textile designer has an infinite source of wondrous and varied surfaces to utilize. A fabric may be rough, soft, smooth, shiny, nubby, coarse, silken, matte, transparent, dull, slick, prickly, or puffy. It may be woven or stretchy, matted, elastic, or furry. It may have a distinct weave such as twill or cotton interlock, or it may be non-woven like felt or synthetic suede. It may or may not have pile, nap or directional sheen. Add color and prints to these textural elements and it is easy to understand why fabric is such a fascinating medium!

I will list some commonly found fabrics and their merits and/or disadvantages. I want to stress, though, that it is important for the artist to experiment with additional materials that are not listed. Do not place restrictions on your creativity!

100% COTTON (CALICO)

This fabric is the one I first learned to love, since I learned to quilt long before I felt comfortable sewing. In fact, for years I relied almost exclusively on calicoes for appliqué, being ignorant and insecure about using other fabrics I knew nothing about. Back then, I was afraid to venture into the unknown; what a shame!

Calicoes do not possess a very interesting texture in their weave – but oh, those wonderful prints! They can be whimsical, sophisticated, old-fashioned, bold, or subtle. I buy cotton prints purely on impulse, for their appeal. I stack the lovely combinations of flowers, pin-dots, feather swirls, paisleys and border prints onto shelves separated by color. Just keeping them highly visible stimulates my senses.

Calicoes are wonderful for adding a three-dimensional feel to a flat surface. For example, a large tree will look more realistic appliquéd in a busy print of greens, browns and golds as opposed to a solid green. The miniature prints are wonderful for appliquéing clothing, wallpaper or patterned furniture, and the larger, more subtle prints are perfect for skies, fields, mountains or roads. When layering many calicoes, be careful to keep a contrast between any two pieces that overlap, either by color or value change, or the total picture may appear too busy. I like to mix and match the large scale with the small scale prints, the solids against the busy prints, the geometric next to floral designs; all the while maintaining a strong combination of light, medium, and dark values.

Calicoes are lightweight, which makes them especially suitable for layering appliqué shapes. Do watch for shadowing, though. If a darker print shows through the lighter print on top, simply fuse interfacing to the wrong side of the lighter fabric to make it opaque. If you prewash your cottons, they become ideal for items that require frequent washing: clothing, baby quilts, pillows, placemats, etc.

FIRMLY WOVEN FABRICS

Includes: poplin, trigger cloth, weaver's cloth, etc. Especially nice as backgrounds, these sturdy fabrics support appliqué shapes and embroidery stitches with little chance of puckering, slippage, tunneling or raveling. They are very useful for beginners as they allow the student to practice maneuvering without the additional concern for controlling the fabric. They are very durable and useful for pillows, furniture covering, duffel bags, or any item which will receive heavy use or frequent washing.

WOOL

Another sturdy fabric, but one that requires more care with washing because the fibers can mat if subjected to a sudden

BELOW: The brilliant colors for the appliqué on this jacket were pulled from the colorful nubs of the wool.

LEFT: The wild jacquard print used to make the pants of this jaunty pant suit provided the vibrant palette for the appliquéd jacket.

ABOVE AND RIGHT: Jeanie Sexton embroidered over printed fabric to create the beautifully shaded flowers in her suit, "Oriental Adornment." The embellished fabric began with a piece cut from the fabric above.

change in wash temperature. For this reason, dry cleaning is preferred.

Wools may be lightweight or heavy, silky smooth or horsehair rough. Whichever, they are absolutely beautiful appliquéd and very manageable. I like to appliqué wool for fur coats on my animals, for clothing when my people are frolicking in the snow, or for furniture coverings or blankets. In other words, appliqué the wool fabric where you would most likely see a woolly texture. Or try appliquéing an entire scene in a variety of wools; the special textural appeal of the wools will create a romantic quality. Wool crepe is one of my favorite fabrics to use in constructing appliquéd clothing; besides being dressy, it has wonderful body to support embellishment techniques.

DENIM

Still another firmly woven, sturdy fabric ideally suited for appliqué. If layering denim or appliquéing to another sturdy fabric, your machine may require a larger sized needle (size 90/14). I prefer the look of "distressed" denim to newly purchased denim; I simply cut up old, cast-off jeans. Denim remains a very "in" fabric to appliqué in large, abstract shapes for teenagers, and it is pretty appliquéd with brightly contrasting thread. Also, denim fringes nicely and will outwear just about any fabric.

POLYESTER/COTTON BLENDS

Any fabric used in dressmak-

ing is perfectly suited for appliqué. Be sure to prewash your fabric first to prevent shrinkage or puckering after it has been appliquéd.

For those who like to sew clothing, a wealth of inspirational patterns can be derived from the fabric itself. Large animal or floral prints are popular. Try cutting around a motif and appliquéing it to a plain background for the top, then use the print to make a companion skirt. You will have an instant coordinated outfit! Many such outfits can be designed using the same principle of incorporating the print itself as the main design. Some polyesters are very sheer, so use the same precaution as with your lightweight cottons and back the sheer fabrics with

LEFT: The bridal satin used to make this striking jacket and gown called for equally glitzy materials for the appliqué: more bridal satin and tissue lamé in 12 different hues.

reach out and touch!

SILK

As smooth as satin but with an additional translucent quality, silk is easier than satin to appliqué, for it does not ravel. Silk changes before one's eyes as the light dances in teasing little bounces. I sense a feeling of movement when I use silk, and it is one of my favorite fabrics. Silk can be expensive, but take my hint to search the thrift shops for old silk clothing. Old silk ties are a treasure to keep under lock and key!

The finer silks should be used only for appliqué shapes as they are too fragile to serve as backgrounds. However, silk noil and the other sturdy silks are perfectly suited for background or appliqué shapes.

POLYESTER SILKIES

These gorgeous, lightweight fabrics are beautiful appliquéd. I use more of the polyester silkies – in wildly patterned prints as well as the intricately textured solid jacquards – than any other fabric. If the texture does not suggest the "mood" I am searching for, the print usually will. So even though I might be appliquéing a rough branch, I can

fusible interfacing to prevent shadowing.

CORDUROY

A rough looking fabric – use it when this rough texture is desired. Try it for tree trunks, fences, sides of a barn or house. The very linear quality of the ribs should be taken into account and positioned accordingly. Try an appliqué shape cut on the bias for a diagonal direction!

SATIN

Aaaaahhhh – didn't the word itself evoke a very pleasurable sensation of softness? Satin can be difficult as it ravels very easily, but it can be tamed with Wonder-Under™. Satin reflects light, which gives it a luminous shine. I use satin when I desire a shiny, smooth effect: ice cream, petals of a flower, balloons, ribbons, etc. Satin is an attention getter and invites the viewer to

TOP & BOTTOM: Polyester silkies were carefully selected to duplicate the texture of the lion fur and the owl feathers for these realistic appliqués.

almost certainly find a jacquard print that is patterned to suggest this texture. The advantage of using silkies for appliqué designs is that the result will be an elegant, dressy combination, the look that I prefer for my clothing.

These exquisite silkies will not be cheap, I am afraid, as they are found in the fine dressmaking department of your fabric store. Fortunately, they are seasonal prints, so I stock up when the prices are slashed during the off-season. Also, a quarter yard cut will make many appliquéd shapes!

RAYON, RAYON BLENDS

Another fine fabric for designing dressy clothing is rayon, which works very well for appliquéd shapes. But most rayon fabrics are very flimsy, so they are not very suitable for using as a background for layered appliqué. But I love rayon, and simply design around the restrictions the drape imposes. For example, couching and couched appliqué or light embroidery would be suitable to embellish rayon. And the market is always coming out with newer blends; some are brocade or tapestry-like, and make marvelous backgrounds. If you are

LEFT: The design for the appliqué on this rayon dress was inspired by the graceful ferns imprinted in the green jacquard fabric.

BOTTOM LEFT & BELOW: This rayon tapestry vest provides a rich surface for swirling Ultra-suede® leaves. The striped jacket pairs nicely with the paisley skirt when the appliqué includes both designs.

uncertain how the fabric will behave, buy ¼ yard cuts and experiment!

METALLICS

I tell my students that I have a lot of gaudiness in my soul, for I cannot accumulate enough of these wonderfully flashy, glittery fabrics. Metallics are no longer limited to gold and silver – one can find just about every hue in this gleaming fabric. Search the bridal and prom section of the fabric stores. In fact, approach your local bridal store and ask who does the alterations. More than likely, they will be only too happy to give you all the scraps you can use!

Metallics can be sporty or extremely dressy and appear very sophisticated. I tend to limit my use of metallics to very simple, bold appliqué shapes because the fabric itself is so flashy that a detailed design would seem lost. Also, the shine can be very intense, so I frequently combine the metallic with another rich looking fabric, such as satin or Ultra-suede®, to "ease" the strain on the eye.

BELOW: A little metallic goes a long way to embellish this rayon tapestry vest.

RIGHT: Clear, strong colors will always look good on a basic black and white plaid. The little red metallic squares fit neatly inside the fabric's print.

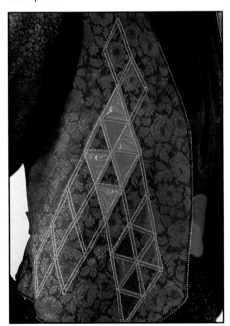

ABOVE: Ultra-suede® provides a soft contrast to the coarsely woven Indian silk of this elegant suit.

SYNTHETIC SUEDE

I consider synthetic suede to be one of the most beautiful, and certainly one of the easiest fabrics to appliqué. Its unique characteristic is the way its cut edge will not fray or ravel. Synthetic suede can be top stitched in place just as successfully with a straight stitch as with a satin stitch.

Synthetic suede will glamorize any garment you apply it to. The rich nature of the fabric itself turns simple geometric shapes or organic motifs into elegant fashion statements. It is extremely durable and washable, but use a press cloth for pressing as a direct iron can leave marks from the steam vents.

One of synthetic suede's disadvantages is that it does not lend itself to multiple layering of appliqué shapes. When I do layer synthetic suede, I combine it with thinner fabrics to minimize the bulk; besides, I like the contrast of the velvety, matte surface of synthetic suede next to a shiny satin or metallic.

VELVETEEN

This is another rich fabric that beckons the viewer to reach out and touch. Velveteen's very plush texture is a natural for fur on animals or clothing for people, but it also makes a bold contemporary statement when used to appliqué shapes not normally considered furry, such as flowers or leaves. Velveteen is a dressy fabric and would make lovely living room sofa pillows or dramatically adorn a beautiful evening gown. It is easy to appliqué; just remember it does have a nap and plan the desired direction accordingly.

COTTON KNITS

Do not be fearful of stretch knit. Its popularity and easy care cannot be denied, and it can be easily stabilized with Wonder-Under™ or fusible interfacing. When I use stretch fabrics as my background, I always fuse either freezer paper or fusible stabilizer to the back to prevent stretching. Save all those knit scraps from sewing projects and you can piece or serge them together to create new yardage. Then cut out a snappy top that will look sensational appliquéd with more knit scraps cut in simple, bold shapes.

SHEER FABRICS: ORGANZA, TWINKLE CLOTH, BATISTE, GAUZE, NETTING, ETC.

What better way to suggest clouds, windows, fog, smoke or ice than with a transparent fabric? Sheer fabrics cannot be backed with Wonder-Under™ so I usually hand baste or pin the appliqué shape in place, since sheer fabrics are normally stiff and easy to sew around. Or, you can free-motion baste the shape in place with your darning foot

ABOVE: Appliquéing cotton knit on top of cotton knit is made possible by the use of stabilizers.

and a straight stitch.

Lovely "shadow" work can be achieved by layering transparent fabrics over brightly colored opaque ones; the result will be soft, muted tones with a romantic appeal.

UPHOLSTERY & DRAPERY FABRICS

I've included the two together, although this covers a wide range. Basically these materials are durable, heavy weight and used in home furnishings. Again, as for other specialty scraps, go to the nearest upholstery business and more than likely they will be glad to "un-load" their leftovers, free of charge.

Machine appliqué is an extremely durable technique and when used in conjunction with more durable types of fabrics, designs can be appliquéd on the living room couch, a favorite arm chair, a tablecloth and drapes with no fear of the appliqué design wearing. It will last as long as the fabric itself. Of course a heavier needle and probably a wider stitch width than usual will be necessary to appliqué through these bulky fabrics.

PAPER

This includes a wide range from barks and rice papers to

RIGHT: This winsome llama was appliquéd with a variety of unorthodox materials: Teddy Bear mohair, Guatemalan cotton dinner napkins, and a wallpaper grasscloth remnant!

handmade papers. Of course, a design made of paper could never be washed, but very beautiful collages can quickly grow under your sewing machine needle. It will be necessary to experiment with the type of satin stitch that will be appropriate, because a close satin stitch could tear a fragile paper. Also, paper dulls needles much more quickly than fabric, but then a very sharp needle is not really necessary to appliqué paper. Just be sure to change to a new needle when returning to work with fabric.

LACE, RIBBON

For the romantic at heart, lace adds that delicate, lacy texture and feminine appeal. Ribbons may be couched, satin stitched, top-stitched or glued. (Use a washable, fabric glue.) Little girls are a joy to adorn in ribbons and lace, as are baby and wedding items.

VINYL, PLASTICS, NYLONS

Very interesting effects can be created with these types of materials, and some are very washable, too. (Be sure to test first.) However, some plastics cannot be stitched with an appliqué or regular foot because the "drag" will not allow the fabric to feed properly. Simply use a darning foot, lower the feed dogs and stitch in place with either a straight stitch or a zigzag stitch.

Vinyl fabrics can be difficult to locate, but there are many excellent mail-order suppliers. Very striking, bold banners and flags can be appliquéd and hung outdoors; the vinyl will withstand inclement weather and decorate the outside of your home for many seasons.

LEATHER

Another specialty fabric that requires a minor adjustment in order to be successfully appliquéd in leather. To work with leather, simply replace the regular needle in your machine with a special leather needle found at most sewing centers or fabric stores.

Leather adds that very unique reptile-like appearance that makes dramatic appliquéd dinosaurs, snakes, alligators, and rhinoceroses for the small boys who are more selective than little girls about what their mothers appliqué on their sweat shirts!

Also, leather combines nicely with rhinestones and dressy fabrics to create a very contemporary look for women's clothing. For example, appliqué large-scale geometrically shaped leather and Ultra-suede® pieces onto a very soft angora sweater. Add a matching leather skirt for a one-of-a-kind garment sure to rate second glances!

HAND PAINTED AND DYED FABRICS

Paints and dyes add a very subtle and mottled surface that makes a perfect background or

BELOW: Judy Simmons demonstrates how beautifully hand-dyed silk lends itself to appliqué in her garment "Light in the Forest."

appliqué shape to highlight with machine embroidery. The dyed fabric combined with embroidery stitches can build very realistic scenes that appear 3-dimensional. Many excellent books can be found for learning how to paint or dye your own fabric. It is an easy and exciting technique to incorporate with your machine art skills. Hand-dyed and painted fabrics can also be found for purchase at quilt symposiums and shows, craft fairs, and specialty shops.

BELOW & BELOW RIGHT: Rosemary Ponte illustrates how she uses the print of a fabric to discover the perfect designs for her wearable art. The printed Dutch-Java cotton blouse fabric was the inspiration for this ensemble.

Again, I want to stress that this is by no means a list of all the fabrics suitable for machine art. Once you have experimented with many different fabrics, you will learn how to control the more troublesome ones by using either Wonder-Under™, fusible stabilizer, special needles, or the darning foot with the feed dogs lowered. Then the whole world of textiles will be open for you to use your sewing machine, machine art skills, and imagination to stitch one-of-a-kind creations!

As for design, many times the texture or print of the fabric itself will inspire the design for your garment or sewing project. Learn to "read into" your fabrics. A hidden nub of color or the

direction of the weave may lead you to discover the perfect pattern for your appliqué. For example, try repeating the check of a print with similar checks appliquéd in contrasting colors or with horizontal bars the same width of the check. A stripe may suggest couching lines, a floral can be cut apart and scattered across a contrasting background or a shiny satin may entice you to machine quilt feathery designs across it. A brightly colored Madras print would be striking as cutwork; simply cut out sections already printed on the fabric and line the piece with a highly contrasting solid. Jacquards are especially inspirational for appliqué designs, with their scrolling, simplified motifs

that can be repeated in the actual appliqué shape with a contrasting fabric. The key is to experiment and have fun!

BELOW: The color scheme is monochromatic, but Rosemary has combined a multitude of unique textures, including 1½" braid used for the petals of the appliquéd flowers.

RIGHT: The jacket's silk lining inspired Rosemary to design these exotic flowers. She sketched the flowers from the print for her pattern and used lustrous silk fabrics to appliqué them to the jacket.

3. Supplies

Today's machine artist can choose from a large selection of tools and supplies to serve his/her needs for creating sensational machine art. Supplies do make a difference, not only in helping the sewing machine perform at its highest efficiency level, but also in making the appliqué and embroidery more beautiful, and faster and easier to create. Manufacturers are well tuned in to machine artists' needs, so there are always new supplies being produced in response to our special requirements.

LEFT: Thread plays critical aesthetic and functional roles in machine art, and can be selected from a wide range of materials.

The following list includes items that I consider essential. Be sure to periodically check with your favorite sewing machine dealer to learn about new products!

THREADS

100% RAYON
MACHINE EMBROIDERY THREAD
(40 WEIGHT)

Technically, good quality dressmaker's thread will work, but do experiment with 100% rayon for a very pleasant surprise. It is an exceptional thread with a very high sheen that seems to transform stitches into lustrous and silky brushstrokes – they become stitchstrokes! Use rayon machine embroidery thread for all your appliqué as well as your embroidery; it comes in a wide range of hues and values and is available on spools or tubes. Sulky™, Mez Alcazar™, Maderia™ and Natesh™ are familiar brands. If you cannot find machine embroidery thread in your area, you may need to rely on mail order suppliers. Be sure to request a thread chart so you can select the colors you want with confidence.

100% COTTON
MACHINE EMBROIDERY THREAD
(50 WEIGHT)

Another very beautiful thread to use for appliqué and embroidery is 100% cotton embroi-

thread, which has a lower luster than rayon thread. This thread is very fine, makes very smooth stitches, and also comes in a very wide range of hues and values. DMC™ embroidery thread, the most frequently available, comes on spools.

METALLIC

You can be sure that I try to get my hands on as many colors of this dazzling thread as I can! Metallic thread will add automatic sparkle to whatever you stitch, and manufacturers are spinning out new colors every season. The variegated ones are my favorites. Some brands are stronger than others, and are less apt to break. You will need to experiment to find the ones you like the best.

SPECIALTY THREADS

These include metallic yarns and ribbon, nubby or variegated knitting yarns, decorative serger threads, Balger™ filament threads, etc. – anything "thread-like" that you love! As you will discover in the following chapters, if a thread or yarn cannot be threaded through the eye of your sewing machine's needle, there are other methods for applying that particular thread. Therefore, I collect threads and yarns with the same undiscriminating passion as I do fabric; if I love it, I buy it – and then find some way to use it!

TRANSPARENT NYLON THREAD

Used for embroidery and couching, this thread is sometimes referred to as "invisible" thread. It comes in two colors: clear to use with light colored fabrics and smoke to use with darks. Request size .004, which is the finest size available and is soft and pliable.

BELOW: Thread becomes the machine artist's palette: over 15 shades of rayon machine embroidery thread were used to shade this rose.

OPPOSITE PAGE
RIGHT: What makes this tropical bird so unique is the variegated rayon thread used to embroider his tail feathers.

LEFT: This time a straight stitch was used to embroider trailing tail feathers in mixed shades of metallic thread.

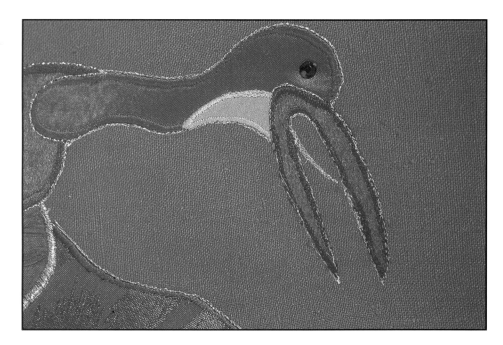

TOP LEFT: A different color of metallic thread was used to appliqué each petal of these roses, and wide metallic ribbon makes a perfect stem.

TOP RIGHT: After this colorful crane was appliquéd, gold metallic thread was free-motion straight stitched around the outline to enhance the rayon satin stitches.

BOTTOM LEFT & RIGHT: Rosemary Ponte uses contrasting thread to accentuate the strong lines of this appliquéd daisy. The close-up shows how the lighter blue thread adds shadows and texture.

OPPOSITE PAGE
TOP: Red metallic yarn enhances this simple fruit motif.

BOTTOM: Brilliantly colored rayon and sparkling metallic threads add luster to this appliquéd and embroidered tropical scene.

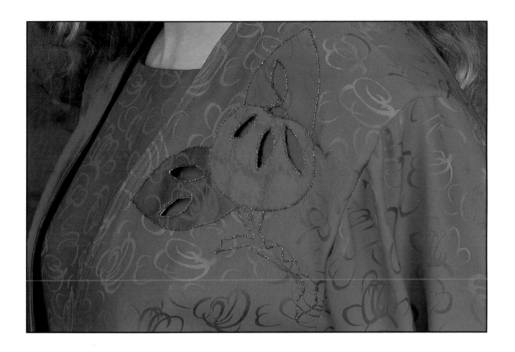

SPECIAL MACHINE FEET

APPLIQUÉ FOOT

This open-toed foot is usually made of clear plastic, and has a groove cut from underneath to allow it to slide over the satin stitches. Your machine may come equipped with one; if not, you should be able to order one to fit your sewing machine, through your machine dealer.

DARNING FOOT

This is probably the most inappropriately named foot in your sewing machine tool box. Does anyone ever darn anymore? Despite its name, this is definitely a handy tool for machine artists. It will allow you to embroider free-motion without having to stretch your fabric in a hoop.

This special foot has a small ring of metal or plastic for the needle to pass through and is fitted with a special spring action to allow it to raise and lower with the needle action. As the needle raises, so does the foot, allowing free movement of the fabric. When the needle lowers, the foot comes down to hold the fabric securely while the stitch is formed. If you did not get a darning foot with your machine, you can order one to fit your particular model.

WONDER-UNDER™

New variations of this product

are appearing throughout the sewing marketplace; it is basically a paper-backed fusible webbing. You will find it in fabric stores among the interfacings and it is usually available by the yard. It is truly a wonder, as it will allow you to accurately trace designs and bond them to background fabric in preparation for appliqué.

OTHER SUPPLIES

LIGHT BOX

A light box is essential for tracing the patterns to be used in machine art. One can be purchased at craft shops or quilt markets or one can easily be rigged up using inexpensive materials. The three things needed are:

• A piece of glass or plexi-glass
• A container
• A light

Look under "glass" in the yellow pages of your phone directory to find your nearest glass supplier. Ask him to cut a square of ¼" thick glass in the size you want (approximately 18" x 18" or larger). Have him sand the edges. Now all you need is something to lay the glass on top of, a simple cardboard box or plastic bin. For your light source, you can purchase an inexpensive "under-the-counter" fluorescent light or use a simple "trouble" light, or any kind of light that will fit inside of your con-

tainer to shine underneath the glass. The whole set-up should cost less than $15, and work as well as the ready-made light boxes that are available at much higher prices.

FREEZER PAPER

You won't find this supply in fabric stores – check your supermarket! Be sure to purchase the polycoated freezer wrap instead of the plastic types. You will bond this to background fabrics – it acts as a perfect stabilizer.

DELI PAPER

This food product supply can be a little more difficult to locate. It is the paper used to wrap sandwiches and hot dogs. Try a paper supply house or restaurant supply house, or ask your butcher. Deli paper is perfect to use as a stabilizer in embroidery because it is a brittle paper and tears away more easily than bond or typing paper. Therefore, embroidery stitches are less likely to tear away with the paper!

FUSIBLE INTERFACING

This product is used as a backing for either very thin or transparent fabrics in appliqué.

TEFLON™ APPLIQUÉ PRESS SHEET™

Fusible products such as Wonder-Under™ and interfacing will not bond to this sheet, so it can be used to protect your iron-

ing board. It is also used to protect your iron when pressing fusible webbing.

WOODEN MACHINE EMBROIDERY HOOP

These sturdy hoops come in many sizes from 4" to 9", and are manufactured especially for machine embroidery. Craft hoops will not tighten enough to hold your fabric secure and are usually too wide to fit under a sewing machine's presser bar.

BLACK PERMANENT MARKER

You'll want the finest point possible. Sharpie™ makes an extra-fine point as well as an ultra-fine point that both work great. This marker is indispensable and is used to trace patterns, mark Wonder-Under™ and draw stitching lines.

WASHABLE MARKERS

These washout markers come in different colors and are perfect to mark on top of fabric for detail lines. Buy one to use on light colored fabrics, and one to mark on dark.

CHALKNER™

Another wonderful marking tool, this plastic applicator has a tiny wheel that rolls along the fabric, dispensing a fine line of chalk as it travels. The chalk can be easily brushed away later when the line is no longer needed.

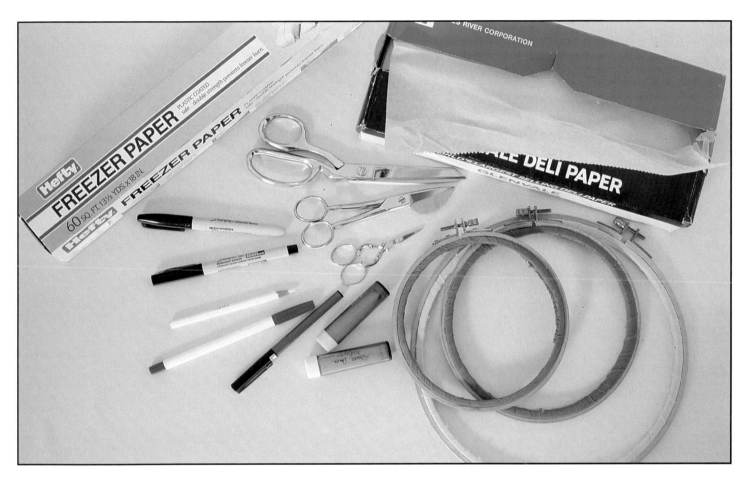

MACHINE APPLIQUÉ SCISSORS

These funny looking scissors have a wide flange on the bottom bill, which allows you to hold them horizontally. Then you can trim right next to satin stitches to expertly clean up any fraying or loose threads.

TWEEZERS

Buy the kind that are manufactured for sewing; the long handle and bent tip make it easy to grasp stray threads for trimming. I also use my tweezers to pull away the stabilizer from the back of my fabric after all stitching is complete.

FRAY CHECK™

This colorless liquid seam sealant is very helpful for machine appliqué.

PERFECT PLEATER™

These fabric-covered louvers enable you to pleat fabric.

If you cannot find most of the supplies listed above at your

ABOVE: Stabilizers, marking tools, scissors and hoops are important supplies for the machine artist.

---

fabric store, you can probably find them at your nearest sewing machine dealer. If they don't have an item in stock, they would probably be glad to order it for you. Sometimes dealers need help in knowing what is in demand!

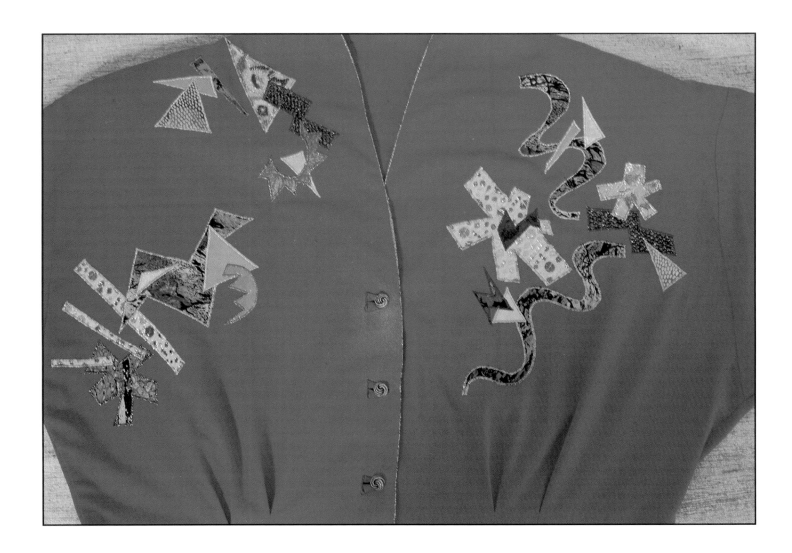

4. Basic Machine Appliqué

I consider machine appliqué to be the "backbone" of machine art. It is the technique I encourage all beginning machine artists to master first. Simply stated, it is layering fabrics and "applying" them to each other, usually with a fine satin stitch. One fabric is used as a background and additional fabrics are appliquéd on top of it. Appliqué is the quickest way to build shapes using the spectrum of colors and intriguing textures found in all those wonderful fabrics you've collected.

LEFT: Basic geometric shapes become elegant embellishments when appliquéd in silky jacquard fabrics using metallic thread.

For best results, select background fabric which is a little heavier, or at least the same weight, as your appliqué fabric. If you appliqué heavy corduroy to lightweight cotton, it is very likely puckering will occur because the heavier material will be on top. It would be more advisable to appliqué the cotton to the corduroy. You may, however, incorporate a wide range of weights in appliqué, or top, fabrics. You may use anything from the sheerest of chiffons to the very thickest of leathers, as long as you keep in mind this one basic rule of thumb: the background fabric should be heavier than the top fabrics.

ADJUSTING YOUR MACHINE'S TENSION

To Start: Place the appliqué foot on your machine. This open-toed foot is usually made of clear plastic and has a groove cut from underneath. This allows the foot to slide over the heavy satin stitches. To test your machine's tension, fill the bobbin with a thread which contrasts in color with the top thread. Place a piece of paper underneath an 8" square of muslin to stabilize it, and you are now ready to test your machine for proper machine art tension.

For basic dressmaking, you must have a balanced tension;

that is, the top tension and the bobbin tension must be equal. For machine art, the bobbin tension should be tighter than the top, so it will pull the top thread to the underside of the fabric. This will result in the smoothest satin stitch because then only the top thread will show on the surface. There is an added bonus: with tensions set this way, there is no need to change bobbin thread to match the top thread. You can use up bobbins leftover from other sewing projects!

Do not be afraid to manipulate your sewing machine tensions! Once you understand exactly what happens when you adjust either the top or the bobbin tension, you will no longer be intimidated by mysterious stitch formations. You will understand what causes stitches to form loops, threads to pull too tightly or too loosely around the fabric, or fabric to pucker. Take a few minutes to practice with your 8" square of muslin, documenting exactly how your sewing machine's tensions work.

Set your machine at its zigzag setting with a medium-wide stitch. As you are stitching, slowly "shorten" the stitch length until a pretty satin stitch forms. Do not make the stitch length too short or stacking will occur. Your sewing machine should feed smoothly on its own, without your pushing or pulling the

fabric along.

Stitch a line of satin stitches and check to see if any of your bobbin is being pulled to the top; this will be easy to distinguish because of the contrasting threads. If you can see your bobbin thread, the top tension is tighter than the bobbin and is, in effect, pulling the bobbin thread up. The simple solution would be to tighten the bobbin, right? However, the top tension is much easier to adjust, so we will loosen the top tension dial to achieve the same effect. The tension is usually very sensitive, so loosen it very slowly as you continue to sew. You will notice that eventually the color of your bobbin thread will disappear, which means that the bobbin tension is gradually becoming tighter than the top tension and

ABOVE: Even a simple flower can be dramatic when appliquéd in snake-skin printed Facille.

is pulling the top thread to the underneath. Be careful not to push the top tension dial too far, or you may notice loops of thread forming on top. These loops mean that the top tension is too loose! No need to panic, simply tighten it up a little!

Occasionally, a student will run into the problem that no matter how far down the top tension setting is lowered, that obstinate bobbin thread still pokes to the top. What to do then? What the sewing machine is saying is that the top tension is still tighter than the bobbin. If you can't lower the top any further, then you must tighten the bobbin.

I remember the first time it

became necessary for me to tighten my bobbin. I was certain all the insides of my sewing machine would become hopelessly jammed. That was because I didn't understand how the bobbin became tighter and what effect it had on the feeding of the thread. I encourage my students to examine their bobbin case, and watch exactly what happens when the bobbin tension is tightened with a screwdriver. If you look at your bobbin case, you will notice an arm directly above the opening where the thread feeds from. When that tiny little screw is turned clockwise, the arm tightens closer to the thread, thus exerting more tension. You will notice this tension as you pull the thread from the bobbin case. When the screw is turned counterclockwise, the arm pulls away from the thread opening, thus exerting less pressure on the thread, making the tension become looser. When I finally grasped the significance tension plays in stitch formations, I felt free to experiment to my heart's content, with my silly "fear" erased forever!

My word of advice to anyone confronted with a sewing machine dealer who warns the customer, "Don't ever touch the tension settings," is to bid the salesperson "Good Day" and leave.

After you have found the correct tension for your sewing machine, record the dial settings for future reference. Settings will vary slightly as a result of differences in fabric weights, but most likely you will stick close to this range for most of your machine art projects.

PREPARING YOUR FABRIC FOR APPLIQUÉ

Now that you have your machine tensions correctly set for appliqué, take another 8" scrap of muslin for background, and some scraps of fabric for appliqué to practice traveling around shapes. Stitching around geometric designs will sharpen your skill for maneuvering. Use the following instructions to prepare your fabric with fusible webbing so you can bond appliqué shapes to the background for appliqué projects.

1. Since we are fusing the fusible webbing to the wrong side of the fabric, it is necessary to use the mirror image of your design for tracing. Place your design over the light box, right side down. Then place the Wonder-Under™, fusible side down, on top of the design.
2. Use a black Sharpie™ to trace the design to the paper side of your Wonder-Under™. Trace each part of the design separately and add ⅛" seam allowance where the shapes overlap. Include all detail lines in your tracing.
3. Roughly cut out each shape traced to the Wonder-Under™, cutting outside the black line.
4. Determine which fabric you want to cut the shape from. Place the Wonder-Under™, fusible side down, on the fabric. I always double check this step, because if the Wonder-Under™ is mistakenly placed fusible side up, it will stick to your iron when you attempt to fuse it. However, you are not fully initiated as a machine artist until you make this mistake at least once!
5. Now you are ready to fuse the Wonder-Under™ to your fabric. Check to make sure that no piece of Wonder-Under™ extends beyond the fabric because any piece that does will fuse to your ironing board cover. Place your iron over the Wonder-Under™ for about five seconds. If you are using heavy or dense fabric, be sure to use the steam setting. It may also be necessary to fuse a few seconds longer at the high setting.
6. Now cut the shapes out along the traced black lines. There is a trick to make it easier to release the Wonder-Under™ from the fabric when you are ready to fuse the pieces to the background. Before you cut along the traced line, peel one corner of the Wonder-

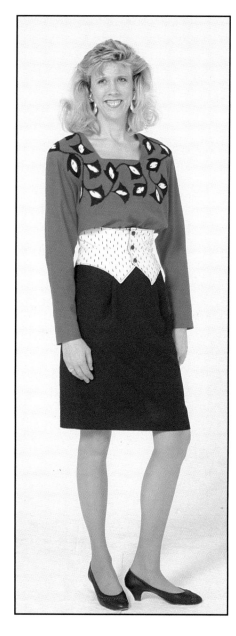

ABOVE: For a coordinated ensemble, use the same fabric for your appliqué as you use for an accessory. The Ultra-suede® used to make the belt for this striking silk dress was used to appliqué the flower petals on the bodice.

Under™ away from your shape. Then, after you have cut out the design, use this corner to peel the paper from the shape.

7. *Before* peeling away the paper, if your design has any detail lines within it, they must be traced to the top of your fabric. Place the shape over the light box and use a water-soluble pen to trace the detail lines to the top.

Use scraps of fabric to prepare shapes on your 8" piece of muslin using the appliqué designs shown below (Fig. 4-1). These geometric designs are ideal to practice maneuvering because they include all the shapes that you will ever encounter in any appliqué project: circles, corners and points. If you travel successfully around all of these shapes, you are ready to appliqué anything!

Your Wonder-Under™ backed fabrics are now ready to fuse to the background. I wait until just before I am ready to fuse before I separate the paper from the appliqué shapes, because they will become tacky once the paper is removed and they may stick to each other.

Peel the paper from the back of your appliqué shapes, starting at the corner you released earlier. Place your background on the ironing board, and arrange the appliqué shapes where you want them. After you are satisfied with the arrangement, press with your iron. Using a press cloth for this step will prevent scorching and iron shine. You are now ready to satin stitch the raw edges.

Figure 4-1

LEFT: Again, simple shapes can be extremely effective, as illustrated in this little girl's dress of 100% cotton. The shapes for the flowers and leaves were cut from folded paper.

SUCCESSFUL MANEUVERING

OUTSIDE CORNERS

Set your machine to the length and tension settings you documented, and use a medium zigzag width. Place a piece of paper underneath the muslin to stabilize it. Move your flywheel by hand until the needle swings to the right, then position the needle just off the outside edge of one side of the square to practice an outside corner.

Satin stitch along that side of your square, with your needle swinging just off the edge of the appliqué. Stitch all the way to the end of the edge and stop with your needle in the down position on the right-hand swing at the dot, as shown in Figure 4-2A. Lift your pressure bar and pivot; then continue stitching to the next corner. This will form what I

refer to as a "squared off" corner, as the stitching overlaps where the direction changes (4-2B).

For most of my corners I prefer to use the "mitered" maneuver, which I believe is much prettier. This maneuver, shown in Fig. 4-3, is only possible if your sewing machine is capable of forming graduated stitch widths.

Start as before, and stitch all the way to the end of the first corner. Stop at the same point, then lift the presser bar and pivot, but not all the way around. You want the very next stitch to dissect the corner like a perfect miter. With a little practice you will learn just how far over to pivot.

Using your hand to move the flywheel, take one stitch until the needle forms a 45-degree angle (A). When you swing the needle to the right, it should hit the very outside corner. Pivot until the

Figure 4-2

A

B

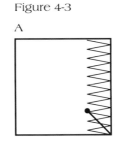

(this area) ▲

Figure 4-3

A

B

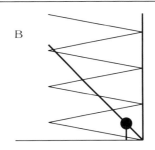

C

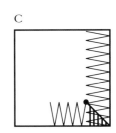

edge you will be sewing along is lined up straight in front of the needle. Now move the flywheel by hand until the next stitch swings over, and decrease your width setting until the needle just hits the mitered stitch (4-3B). Next, you will gradually increase your width setting until the stitches pass the mitered stitch; at this point the width should be the same setting that you started with (4-3C). This mitered corner stitch takes a little practice, but is worth the effort to master as it finishes the corner much more smoothly.

TIP: If your sewing machine has the "needle-down" feature, your maneuvering will be dramatically simplified, as you will not ever have to hand-control your flywheel. Start your stitching with the "needle-down" button pressed, so that when you reach a corner or a point and stop at the end, the needle will automatically stop in the down position. After you pivot, press the "needle-down" button and the needle will raise up. (This is easier than moving your fly-wheel by hand to raise the needle.) Now lift the presser bar and position your fabric below the needle, exactly where you want it. You may also re-adjust stitch width for a taper or miter at this point. Then lower your presser bar, re-engage your "needle down" button, and continue. Again, this maneuver will take a little practice, but will save you considerable time and effort. I tell my students that they will want to kiss their "needle-down" buttons after working with machine appliqué for a short while!

INSIDE CORNERS

When approaching an inside corner, stitch all the way to the bottom of the edge. From this point (Fig. 4-4A) you will need to continue stitching for the distance of your stitch width; then stop with your needle swinging on the inside of your appliqué shape (Fig. 4-4B). Pivot, then swing the flywheel by hand until the needle swings over to the right and down. Scoot your fabric, if necessary, until the needle just hits the stitching; then continue (Fig. 4-4C).

OUTSIDE POINTS

For a blunt point, stitch as for a corner, stopping with your needle at the outside tip of the point (Fig. 4-5A). Pivot, then swing the flywheel by hand until the needle swings over to the left and down; you will notice the needle about to enter the background fabric. Lift your presser bar and scoot your fabric until the needle is just above your stitching (Fig. 4-5B). With the blunt point maneuver, your point will always be as wide as your zigzag width setting.

Again, a prettier, tapered point is possible only if your sewing machine is capable of forming graduated stitch widths. This time, slowly approach the point until you notice that your next needle swing is about to hit the background (Fig. 4-6A). Now decrease your stitch widths and bring your needle over to the right. Take one or two more stitches until you notice that the needle swing is ready to hit the

Figure 4-4

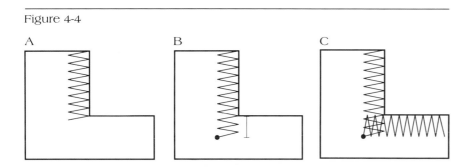

Figure 4-5

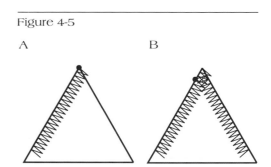

background again. Decrease your stitch width and continue in this manner until you reach the bottom point (Fig. 4-6B). However, I do not like to decrease my stitch width below 10mm, as my needle seems to just poke holes if it is narrower than that. With the needle down on an outside swing, lift the presser bar and pivot your fabric, in preparation for stitching the other side of the point. Turn your flywheel by hand until the needle swings over and down. (This should be becoming very automatic by now!) Scoot your fabric until the needle just hits the stitching and take a few stitches. Continue stitching, gradually increasing your width to the original setting. Your stitching should be just

leaving the mitered stitches. You will have formed a perfect miter, and be able to create very beautiful, tapered points in this manner (Fig. 4-6C).

INSIDE POINTS

An inside point is formed in much the same manner as the inside corner: you will stitch all the way to the bottom of the edge and then continue stitching for a distance equal to your stitch width (Fig. 4-7A). Stop with the needle swinging on the inside of your appliqué shape and pivot. Turn the flywheel by hand until the needle swings over to the right and down. Scoot your fabric until the needle just hits the stitching; then lower your pressure bar and continue (Fig. 4-7B).

You may also taper an inside point, by again utilizing graduated stitch widths. As you approach the inside point, stitch all the way to the bottom of the edge and stop with your needle on the inside swing. At this point, I find it helpful to visualize a straight line at the center of the miter (Figure 4-8A). Bring your needle to the right and decrease your stitch width until your needle hits where the miter would be. Continue in this manner, increasing your stitch width as needed to form the miter. As you pass the mitered stitch, your width setting should be the same that you started with (Fig. 4-8B).

OUTSIDE CURVES

The trick for satin stitching perfectly smooth, rounded curves is to take your time and pivot often. Many students try to push their fabric around a curve and the result is uneven, angled stitches. The tighter the curve, the more often you will need to pivot; for the tiniest curves it may be necessary to pivot after every single stitch. How do you know when you should pivot? The stitches should aim directly into the appliqué shape from the edge. When you notice the stitches starting to angle, it is time to pivot. You will easily know when you are approaching corners and points, but you will need to train your eye to

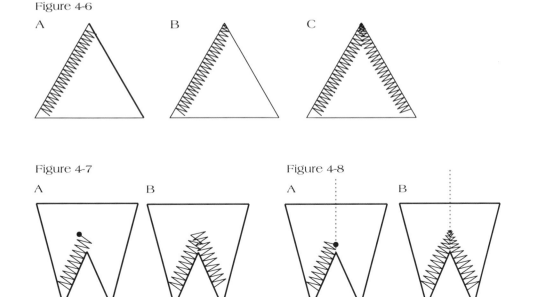

Figure 4-6

A   B   C

Figure 4-7

A   B

Figure 4-8

A   B

know how often and when to pivot on curves. This will take practice, but after a while you will find yourself rounding curves very automatically.

Travel around the outside curves clockwise and always pivot with the needle on the right-hand, or outside swing (Fig. 4-9). If it is necessary to stitch counterclockwise, just remember to keep your needle on the outside (this time the left-hand) position. Do not pivot too sharply or your next stitch will fall off the appliqué edge. Remember, take your time!

INSIDE CURVES

The same procedure applies to inside curves; that is, pivot when the stitches start to angle off the edge of your appliqué shape. But for inside curves,

Figure 4-9

Figure 4-10

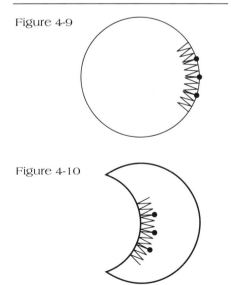

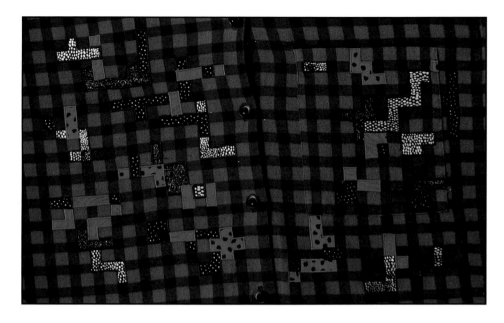

ABOVE: Closely inspect your fabric if you find yourself stumped for an appliqué design: chances are the perfect solution will emerge!

pivot with your needle on the inside swing (Fig. 4-10).

These rules are easy to remember because for every outside maneuver, whether it is a corner, point or curve, the pivot point will be on the outside. Likewise, for all inside maneuvers, the pivot point will be on the inside. The steps are easy to learn, but no one can become an accomplished appliqué artist without spending some time practicing! The maneuvers will seem awkward at first; just remember not to expect perfect results with your first practice piece. I promise that if you do practice, after a very short time you will be very proud of your beautiful stitching – and best of all, it will start to feel very automatic and easy!

TYING OFF

When you have completely traveled around a shape and end up where you started stitching, it is necessary to *tie off* so that your stitching will not come unraveled.

Slowly approach your beginning stitches and travel right up to where they started. Make sure that your stitches are parallel to each other. If they are not, take a few more stitches, pivoting when necessary, until they are lined up. Now, stitch over your first stitches by several threads. If you own a computerized sewing machine with the *tie off* feature, you simply press the correct button and your machine does the rest. If not, you will need to manually tie off. First, bring your needle to the up position; then change to a straight stitch. When you do this, you will notice that the needle centers itself. Since you do not want to stitch right in the middle of the pretty satin stitches, lift the pressure bar and scoot your fabric until the needle is positioned right at the edge of one side of your appliqué stitches. Now take three or four stitches with the straight stitch; your stitch length should be set so short that the stitches will seem to fall on top of each other. This forms a knot on the underside and effectively keeps your stitching from unraveling. Cut your top thread close, and on the underside clip the thread just beyond the small chain of knots that you created.

ACCURATE POSITIONING OF APPLIQUÉ SHAPES

After you have practiced maneuvering, you will doubtlessly become very excited about putting into practice your newly acquired skills! You will find yourself evaluating the various shapes you appliqué as inside corners, outside curves, tapered points, etc. The more experienced you become, the easier each project will be.

Soon, you won't even consider the appliqué a challenge. But what will continue to be a challenge with each new project will be the selection and evaluation of the designs and the choosing of perfect fabrics.

As you progress to more complicated designs, it will also become critical to place each appliqué shape exactly, according to your drawing. The larger the number of pieces that overlap and the smaller the pieces become, the more difficult it will be to "eyeball" placement for your shapes. I have devised a simple and foolproof way to place appliqué shapes exactly as they are in my original drawing.

1. Place the drawing right side down on a light box.
2. Cut a piece of freezer paper the same size as the drawing and place it, fusible side (shiny side) down, on top of the drawing on the light box. Trace the drawing using a Sharpie™. You will have a mirror tracing of your image.
3. Refer to the "Preparing your Fabric for Appliqué" instructions to prepare your appliqué shapes with Wonder-Under™. Start with step 2 and use the freezer paper tracing to trace each shape, since it is the mirror image.
4. Place the freezer paper, fusible side down, against the wrong side of the background fabric. Press with a hot, dry

iron until the paper bonds to the fabric, approximately 8 seconds. Do not slide the iron; rather pick it up to move it to another area.

5. Place your background, right side up, on the light box. Your placement lines will show through so that you will know exactly where to place each appliqué shape. Arrange your appliqué shapes on the background, overlapping them as necessary. Carefully carry everything to the ironing board, pinning shapes if necessary to prevent them from slipping, and fuse all into place.
6. Your freezer paper will also serve as a stabilizer! If the edges come loose as you are appliquéing, re-fuse them. Tear away all paper when the sewing is complete.

TIP: I have found that if I place a ½" thick sheet of foam rubber between my light box and my background, I can fuse my appliqué shapes directly on my light box. The image will easily show through the foam and I am assured that none of the pieces will slip out of position before fusing. "Tack" the appliqué shapes, without pressing hard or long with your iron, to prevent your glass from becoming too hot. After all shapes are tacked in place, carry the background

to the ironing board for a more thorough fusing.

This method for transferring a design will work for the majority of your projects. However, if you are using a heavy or dense background fabric, such as corduroy or Ultra-suede®, the placement lines will not show through the light box. If you are appliquéing a ready made sweat shirt, you would probably find it too cumbersome to place over the light box. In either case, an easy alternative to the freezer paper method would be to trace your design onto a piece of tracing paper using a black permanent marker. Then place this tracing over your background, centering it where you want it. Pin it in place with just two straight pins at the top. Then flip this tracing up and down, placing each appliqué shape in position underneath the traced image. You can fuse the shapes in position directly through the tracing paper, either as you progress, or after the entire image has been arranged.

LEFT: The jacket pattern pieces for "The Beast Within Me" were cut from freezer paper, and basted together before the design was drawn on. Then the basting was removed and the freezer wrap was bonded to the wrong side of the green silk. This provided the placement lines for the appliqué pieces, which then flowed smoothly across the seams. (See page 49.)

APPLIQUÉ TIPS

- Test fabrics with any unusual content for fusing before applying to your background. Fabrics needing testing include: metallics, sheer fabrics, acetate satins, man-made leathers, brocades. You want to know exactly how the fabric will react to a particular heat setting; adjust accordingly.

- Test any lightweight or light colored fabric for shadowing when layering; if a darker fabric underneath it shows through, then back the lighter fabric with fusible interfacing to make it more opaque.

- Be careful when layering heavy fabrics; they may become too bulky. You want to avoid ridge marks caused from fusing a lightweight fabric in a position where it overlaps a heavier fabric.

- Normally you will match the thread as closely as possible to color of each appliqué shape. This will make your shapes appear more three-dimensional. If you use contrasting thread, the shapes will appear to be outlined.

- Using a narrow stitch width to appliqué (approximately 2mm), will make it easier to maneuver! It will also necessitate fewer steps to taper, and allow the foot to glide more easily over corners. Try it!

- It is also easier to appliqué straight edges if you press your foot pedal to sew fast! When traveling slowly, you will tend to jerk or wobble, or pull or push your stitching along. When stitching at a fast pace, the machine does the feeding and your movement will automatically become smoother. Again, try it!

- Your order of stitching appliqué shapes is important. Start with the back (underneath) appliqué shapes first, working in sequence to the shapes on top. Begin stitching a few threads inside any adjoining shape and end your line of stitching slightly into any adjoining shape (Fig. 4-11). Your subsequent satin stitching will neatly cover these thread ends.

- Tweezers are wonderful, not only for cleaning up frayed edges, but also for removing the stabilizer from the back of your fabric when the appliqué is complete.

OPTIONAL EDGE FINISHES

Satin stitching the raw edges is certainly the most popular technique for machine appliqué, but there are other options. By lengthening the stitch dial, you can get a longer stitch that will ultimately become a regular zigzag. This may be more appropriate for some materials, for example nylon or vinyl. Or perhaps you want to create a striking graphic design using highly contrasting thread: a zigzag stitch may add the perfect visual impact as an additional design element. Just be aware that the farther apart the stitches are spaced, the more susceptible the edges will be to wear. Choose your materials with this in mind. If you zigzag lightweight cotton, for example, the edges will most probably fray with extended handling or washing. On the other hand, Ultra-suede® can be appliquéd with a very long zigzag stitch (or even a

Figure 4-11

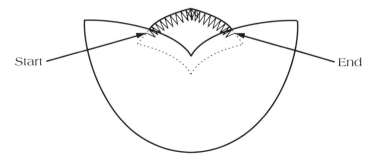

Start   End

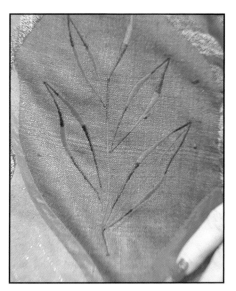

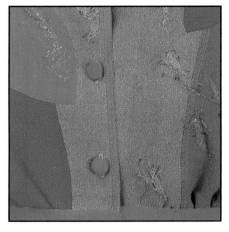

straight stitch) and will not fray ever! And if you desire that hairy, frazzled look, you can deliberately choose fabrics that fray easily, such as loosely woven linens, wools and metallics, and appliqué them with a long, wide zigzag setting using metallic thread. The effect will emphasize the highly textured surfaces of your fabrics.

Also, if you own a computerized machine, many of the decorative embroidery stitches can be used to appliqué. It will be necessary for you to experiment with your own stitches to achieve the effect you desire. Be sure to practice and explore how the stitch looks on the edge, and also how easy it is to maneuver around corners and curves. For this reason, when I plan to use computer stitches to appliqué, I choose simple designs for my appliqué, designs with shallow curves and fat points and corners. But computer stitches can make even simple designs appear very exciting!

TOP LEFT: Oversize poppies appliquéd in rust taffeta: gorgeous!

TOP RIGHT: Rosemary Ponte has turned a simple leaf motif into an elegant design by manipulating the width settings of her satin stitch to form a tapering flow of variegated thread.

ABOVE: A computer stitch was used to finish the raw edges of these bold abstract shapes.

MORE APPLIQUÉ
SHAPES TO HAVE
FUN WITH

5. Couching

One of my favorite edge finishes is to simply lay a pretty cord, such as a metallic yarn, along the outline of the appliqué and zigzag it in place. This is called "couching" and it is extremely beautiful. As I combine a variety of couching materials (narrow ribbons, decorative serger threads, interesting knitting yarns, etc.) with decorative threads threaded in the machine (rayon or silk, variegated, metallic, etc.), and then use various settings to stitch (zigzag or simple computer stitches), the variables become limitless! Best of all, this is a simple technique that does not require much practice. The secret is to choose glitzy materials and simplified designs – such as the overlapping diamonds in the exercise beginning on page 56.

LEFT: Couching over metallic yarn beautifully finishes the raw edges of these strips of brilliantly colored fabrics, which were simply fused in place.

RIGHT: Gold metallic yarn was used to outline couch the blue river in "Whoop It Up." It heightens contrast, making the river stand out against the blue silk background.

DRAMATIC DIAMONDS

First, select the material for your background, and cut a 16" square. Just about any fabric will be appropriate – even stretchy knits or lightweight silk or rayon. Since we are not satin stitching the edges, we will not have to be too concerned with using a firmly woven background. This is why I use this technique frequently when I want a spectacular, dazzling design for a garment made from fabric with a fluid drape (for example, rayon or polyester) or a tendency to stretch (cotton or jersey knit). If I appliquéd using the more rigid satin stitch, my fabric might lose

LEFT & BELOW: Dramatic diamonds appliquéd in shiny jacquards and metallics bedazzle this green silk suit, "Trailing Stars."

its lovely drape or stretch out of shape. I always test the design using the actual background and appliqué fabrics as well as the thread and couching material I plan to use. Every fabric combination will respond differently, and I'd much rather waste 10" of expensive charmeuse as a test scrap than the front section of a cut-out blouse!

You will need to be more selective for your appliqué fabrics. They must be lightweight and not tend to ravel. My favorites for this technique are those gorgeous polyester silkies; they are very thin and do not ravel easily, plus are wonderfully brilliant in a multitude of colors, patterns, and textures. Again, experiment with your own favorite fabrics.

After you have selected your appliqué fabrics, trace the diamond pattern, (Fig. 5-1) on a piece of template plastic and cut it out. This is the only pattern you will need to create *hundreds* of different designs!

Trace about 12 of these diamonds onto the paper-backed side of a piece of Wonder-Under™, leaving about ¼" between diamonds. Then cut each one out separately, cutting just beyond the traced line. Next, fuse the diamonds onto the wrong side of your appliqué fabrics, fusing several of them on your favorites. This technique allows you to incorporate a great

many different fabrics. I have designed whole jackets using a different fabric for each diamond – my own version of "charm" clothing! On the other hand, a design using only two or three different fabrics can be just as dramatic. You are the designer – the fun is selecting the fabrics that appeal to you the most!

Figure 5-1

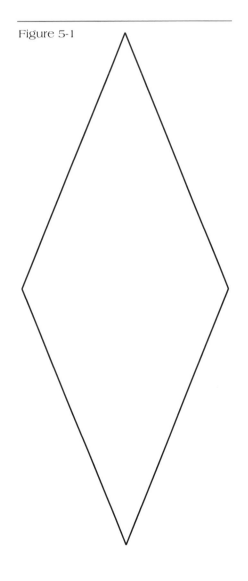

Cut your diamonds out on the traced line, remembering to peel up one edge first for easy removal later. Now spread out your beautiful diamonds, enjoy their sparkle for a few moments, and reserve them for later. Do not be concerned with where they will go, how many you will use or which ones will lie next to each other. Be patient. Those decisions will come later, and will be made spontaneously.

Now focus your attention on the background fabric. You will need a 2" clear plastic ruler and a Chalkner™. Place your ruler diagonally (on any diagonal – the angle is not important). With your Chalkner™, mark a line on both sides of the ruler, dissecting your block. This 2" grid will form the basis for the foundation of your design. Now select one of your diamonds and remove the paper backing. Place it inside the grid with the long edges of the diamond directly on the marked

lines. This diamond will be the "cornerstone" from which you will build your design (see photo bottom left). Select another diamond and carefully line it up next to the first diamond; it may be placed above, beneath or on either side. Just make sure its edges barely touch those of the original diamond and that the points meet perfectly. Now the fun begins! Continue to build your design, making decisions one at a time. Remove the paper backing each time you place a diamond, but do not fuse the diamonds in place until you are satisfied with the entire design.

Your diamonds will have a very definite one way direction (see photo bottom right) if you

BOTTOM LEFT: The "cornerstone" diamond from which to build your design.

BOTTOM RIGHT: This arrangement of diamonds gives a definite one way direction.

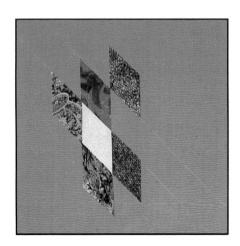

continue to place them with the long edges parallel to each other. If you want a less static design, you can easily add another diagonal direction by reversing the diamonds so that the short edges touch to form a chevron. This subtly adds another diagonal running in the opposite direction, resulting in a very pleasing combination of shapes (as shown in photo below)). You can even continue the diamonds to form an eight-pointed star! The possibilities are limitless; it is very easy to build your own spectacular original design.

BELOW: Form a chevron with your diamonds to add another diagonal running in the opposite direction.

RIGHT: Triangles are cut from polyester and silk, then bonded to wool crepe in "Timeless Shadows." Couching neatly finishes the raw edges.

FAR RIGHT: Detail, "Timeless Shadows." Triangles are stacked to form a pyramid of color and texture.

Once you are pleased with your combination, carefully fuse the diamonds in place. It is important that all edges and points meet, so fuse just one or two diamonds at a time to make sure they do not slip out of place. Your design is pretty enough as it is, but we are going to add more textural embellishment with the couching that finishes the edges.

Your couching material may be any kind of yarn, ribbon or cord that is basically flat and no wider than ¼". My favorites are the gorgeous decorative serger threads on the market, including Candlelight™ yarn and ribbon floss. Pretty knitting yarns can also be used, as well as other ribbons and craft yarns and cords. After you get "hooked" on

couching (and you will!), you will start to notice all kinds of appropriate couching materials on the market, especially at craft and quilt shops and venders' booths at quilt shows.

There are more decisions to be made, as you contemplate what to thread your machine with. You will find that rayon machine embroidery thread is an ideal choice and is much more attractive than regular thread. I also love to use metallic threads in all colors, and also variegated threads. If you use clear nylon thread, your stitching will be virtually invisible so that your couching yarn will seem to be appliquéd by hand. This is a perfect choice for a yarn or ribbon that is so beautiful you don't want any of it covered. And if

you are able to use a twin needle in your sewing machine, you will be able to use two colors of thread (perhaps one metallic and one variegated rayon) to create an exciting stitch to complement your couching yarn. As for color – sometimes I choose a thread that will blend with whatever I am using for yarn, and other times I like to select a contrasting thread. Your bobbin thread will not show, so you can use any color of regular dressmaking thread.

The most commonly used stitch will be the zigzag; set to the length of around 10-12 stitches per inch. As for stitch width, you will want the needle to just clear whatever you are couching. The needle should fall just off the edge of your yarn on either side. Therefore, you will use a narrow width for a narrow yarn and wider widths for wider yarns. You can use either an appliqué foot or an all purpose foot. Place a piece of tracing paper or "deli" paper (see "Supplies") underneath for stabilizer.

To start, lay the couching yarn along one edge of the fused diamond and position the needle directly above the starting edge. You will need to put about 2" of yarn to the back of your foot to be able to hold it firmly in place. Take one stitch, reverse one stitch, and then continue stitching, zigzagging the yarn in place. Be careful that the

yarn hugs the edge of the fused diamond, to cover the raw edges. After a few stitches, you will be able to hold the yarn tautly in front of the foot as you progress. When you reach the end of the edge, press your tie-off button if you have one. If not, backstitch one stitch and then set your length to "0" and take two or three stitches in place. Cut the top thread and the yarn close to the fabric and cut the bobbin thread from underneath. Then trim the yarn at the start, even with the edge of the dia-

mond. Another method of tying off doesn't involve changing the stitch length. Simply press the fabric to the throat plate to hold it in place while you take two or three straight stitches, then clip the thread and yarn. This will effectively tie off the end so that it will not come unraveled.

I always test my couching materials before I begin on an actual project. I like to experiment with different yarn, stitch

BELOW: A sampler of couching yarn/stitch/thread combinations.

ABOVE: Very little appliqué and a lot of couching form crisp diamonds in "Diamond Icicle."

BOTTOM LEFT: Dramatic Diamonds with the couching added. Notice that some couched lines extend beyond the fused diamonds.

BOTTOM RIGHT: Jewel-toned squares and tiny strips of glittery fabrics make this knit top special when outline couched with metallic yarns.

and thread combinations on a scrap, and then I keep this beside me for reference. I can tell at a glance what a particular thread looks like against two or three different yarns, and perhaps vary the stitch by substituting a simple computer stitch instead of the zigzag, or using a twin needle, etc. I keep these swatches; they save me considerable time when I am ready to work on a project in the future. The technique is very simple, but the decisions that you are faced with can be time consuming. Once you start your project, these sample swatches significantly reduce the time you spend making decisions.

Before you begin couching your design, analyze it, and start couching the shortest lines first. In other words, you want to make as many continuous lines as possible, preferably so that

these longer lines cross over the shorter lines underneath, to cover the raw edges of your yarn. However, you will discover that this is not always possible and many times it will be necessary to end your yarn against a line that has already been couched. When this happens, tie off as usual, and then add a dab of Fray Check™ at the intersecting yarns before trimming close.

You can add additional lines, drawn with your Chalkner™, to extend couching beyond the fused shapes (see photo bottom left). You can also add couching inside appliqué shapes (for example, the veins inside a leaf), and you can couch along the outside of satin stitching. You will find more and more ways to incorporate couching in your designs to add a very beautiful and tactile quality to your machine art.

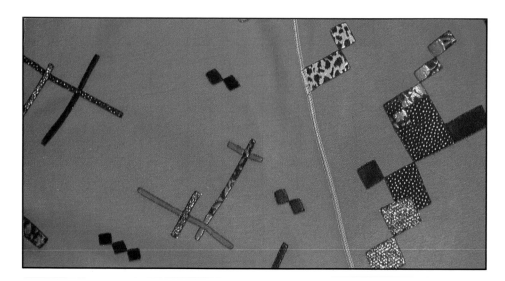

COUCHED SEMINOLE

I love crisp, geometric patterns, and Seminole patchwork has always been one of my favorite piecing designs. However, I become frustrated by the narrow seams and multiple seam allowances involved in strip piecing Seminole patterns. Also, I like to use silky fabrics in my designs and these can be very slippery to machine piece. So I appliqué my geometric patterns instead, by fusing strips of fabric in position, much as the diamonds were appliquéd in the previous exercise.

Templates aren't necessary because a rotary cutter is used to slice foundation strips of fabric into various widths to be arranged on a background. Smaller rectangles and squares of fabric are cut to form checkerboards and stripes, and these are arranged on top of the foundation strips. The underneath fabric works as a negative design to fill in the checks and stripes, so that the resulting pattern seems much more intricate than it really is, and the raw edges are couched with metallic yarn to add the final pizazz.

RIGHT: The same fabrics used to appliqué the owls in "Spirits of the Night Wind" are used to cover the cape-like collar with couched Seminole. The yarn selected for couching subtly blends with the muted hues of the Seminole couched fabrics.

MATERIALS:
Rotary cutter, mat and ruler
½ yard Wonder-Under™
15" square background fabric
¼ yard of four to six fabrics for
    appliqué
Chalkner™

To start, cut at least four strips of Wonder-Under™ approximately 22" x 5"; then fuse them to the wrong side of 4 different fabrics and cut around the Wonder-Under™. Place each strip on the rotary mat, and with your rotary cutter, slice it into a combination of widths ranging from ½" to 3". I place my fabric on the mat with the paper side up so that I am sure that the rotary blade cuts through both the paper and the fabric. Remove the paper backing from each strip of fabric.

With a Chalkner™, draw a diagonal line across the background fabric, and choose one of your wider strips for the foundation. Place this strip along the chalk marked line, with the ends extending beyond the edges of the background fabric. Carefully fuse it in place and trim the ends even with the edge of the background.

With your Chalkner™, draw another line perpendicular to the first one. With your rotary ruler and cutter, square off the end of the next foundation strip you plan to use, so that it forms a perfect right angle. Place this

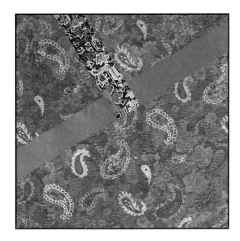

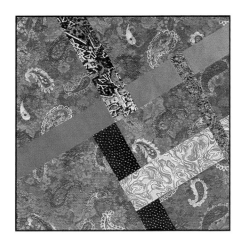

strip along the chalk marked line, with the squared end even with the edge of the first strip, and fuse it in place (photo top left).

Add more strips in different widths, overlapping and weaving them next to each other, as shown in the photo top right. I like to draw my lines first with my Chalkner™, to make sure they are exactly straight and perpendicular to each other. After you

are pleased with the basic arrangement of your foundation strips, carefully fuse them into place. You can then fuse smaller rectangles and squares of fabric to form checkerboards and stripes on top of the foundation strips, as shown in the photo bottom left. These can be hand cut or sliced with the rotary cutter.

After the design has been bonded to the background, couch the raw edges of your

OPPOSITE PAGE
TOP LEFT: Two strips of fabric are placed perpendicular to each other to form the foundation.

TOP RIGHT: More strips in different widths are added; overlapping and woven next to each other.

BOTTOM LEFT: Smaller rectangles and squares of fabric are added to form checkerboards and stripes on top of the foundation strips.

BOTTOM RIGHT: Complete design with the couching added.

fabric strips with metallic yarn (refer to "Couching"). You may wish to place a piece of batting underneath the background to quilt your design as you couch. Additional lines may be drawn with your Chalkner™ to extend your couching beyond the fused strips (photo bottom right, page 62).

It is very exciting to see how quickly a beautiful geometric pattern will emerge using this technique. What I like most, is that this technique does not entail a lot of pre-planning. The strips evolve very rhythmically and spontaneously and are easily adaptable to uneven background spaces, which makes the technique especially suited for wearables. The only disadvantage of building your designs intuitively is that you are likely to end up with a collection of left-over rotary cut strips. Simply store them in a shoebox for your next project!

ABOVE: The bottom band of this tiered skirt is patterned with couched Seminole of hand-dyed cottons. The jacket has also been couched in swirling loops in a variegated nubby knitting yarn.

COUCHED QUILT BLOCKS

Traditional quilt patterns are timeless, and contemporary artists will probably always use them for inspiration. I like to incorporate machine art and unconventional materials and methods to interpret favorite traditional quilt patterns. Instead of piecing or appliquéing my blocks in cotton, I bond my designs using glitzy metallic and jacquard fabrics, then couch over the raw edges using beautiful yarns. Instead of hand quilting the background, I again incorporate couching – this time through a layer of batting. I like to trace old fashioned quilting patterns such as feather motifs and ½" linear grids with my couching yarn. Sometimes I add further texture by inserting tiny pleated fabric into parts of the quilt blocks pattern.

The techniques that I use to modernize quilt blocks are very similar to those used for the Dramatic Diamonds and Couched Seminole. However, instead of using a single template or the rotary cutter to create the design, appliqué shapes are traced from quilt patterns to Wonder-Under™. The Wonder-Under™ is then bonded to the appliqué fabrics, which are fused to the background to form the block.

Perhaps you have a collection of quilt books from which to

select a pattern to use. Either pieced or appliqué designs are appropriate, in any size from miniature to over-sized. Most books include full-size templates for their patterns but usually include only a small scale diagram of the finished block. You will need an actual size draft of your finished block, so the first step will be to draw a square that is the exact size of your finished block. Trace each template (do not include seam allowances) onto a piece of template plastic and cut it out. Refer to your block and position each template inside the square to trace an actual size draft of your block.

Trace this pattern onto a piece of freezer paper, to be bonded to the wrong side of your background fabric. This will serve to show the placement lines for your appliquéd shapes.

Remember, if your pattern is asymmetrical, you will need to use the light box to trace a mirror image of your pattern (refer to page 38).

ANALYZING THE BLOCK

Before you start tracing shapes onto your Wonder-Under™, analyze your block to determine whether it might be simplified. If your block has a pieced background, you will be able to simplify your pattern. For example, when I made Carolina

Lily, shown on page 64, I was able to eliminate the pieced background and substitute a solid square that was the same size as the finished block.

Also, if your design has many tiny units pieced next to each other, you can fuse one whole shape to serve as the foundation and simply appliqué the small units on top. This involves the same principle used in Couched Seminole, and is

much simpler than trying to fuse tiny shapes right next to each other. Look at the photograph, bottom left, of Indian Wedding Ring before the little Sawtooth units were added. You can see that the double rings were simplified and fused as whole units first. In the photograph to the right, you can see how the little red teeth were fused on top of the rings, to give the appearance of a pieced design.

OPPOSITE PAGE: This Carolina Lily block is based on the traditional pattern, but a solid background square has been substituted for the pieced background.

BELOW: This block is based on the traditional Indian Wedding Ring design. The double rings were simplified and fused as whole units first.

RIGHT: Here the triangle units have been added on top of the solid rings, giving the appearance of a pieced design.

ADDING PLEATED INSERTS

The little pleated sections in the Indian Wedding Ring block are added using reverse appliqué; in other words, the pleats are placed behind the background and then that layer is cut away to reveal them. This is necessary because pleats form very bulky edges that would be difficult to appliqué down. When the background is appliquéd to the pleats, the edges of the pleats are smoothly covered by the background fabric. This technique can be used to appliqué other troublesome fabrics such as metallics that ravel easily and heavy brocades.

First, decide in which sections you want to include pleated inserts, and then determine the direction that you want the pleats to run within that section. You will need the Perfect Pleater™ by Clotilde® to form the pleats (see "Suppliers" at the back of the book). Follow the instructions included with your pleater to form crisp, narrow tucks out of the fabric you have selected for the pleated sections. You will need to cut your fabric four times the length of the section to compensate for the take-up required to form pleats. For the width, simply add two inches, no matter what the size. If you are pleating an area to fit inside a 3" square, for example, you will need to cut your fabric 5" wide and 12" long. After you have pleated the fabric, it will be approximately 5" x 5". I prefer to prepare one section at a time rather than pleat yardage to cut my patterns from. Small pieces of fabric are very simple to manipulate in the pleater, and there is less waste of pleated fabric.

Tuck your fabric into the pleater, right side down. After you have pleated your fabric, bond fusible interfacing to the wrong side of the pleats before

STEP 1

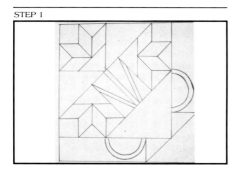

STEP 2

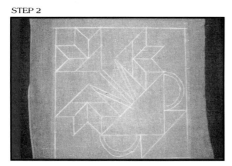

STEP 3

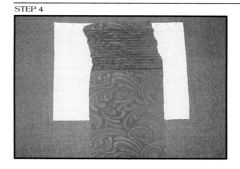

STEP 4

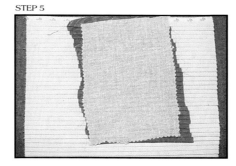

STEP 5

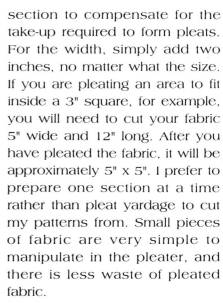

STEP 1: Quilt block traced full size on freezer paper and fused to the wrong side of the background fabric. Image was reversed, and traced with a Sharpie™.

STEP 2: Placed over a light box, the image clearly shows through and is traced with a Chalkner™.

STEP 3: Appliqué pieces are positioned over the placement lines and fused in place. The spaces left open are designated for pleated inserts.

STEP 4: Using the Perfect Pleater™, pieces of fabric are tucked, right side down, to form pleats.

STEP 5: A piece of fusible interfacing is bonded to the fabric while it is still tucked inside the pleater, to permanently set the pleats. The fabric is then removed.

removing the fabric from the pleater, following instructions included in the pleater. Use the light box to position the pleats underneath the background, exactly where you want them inserted. Pin the pleats in position, being careful to keep them running straight rather than at an angle. Next, straight stitch directly on the outline of the section, using a very short stitch length (approximately 15-18 stitches per inch) and matching thread. Stitch completely around the section, pivoting when necessary.

Now use your machine appliqué scissors to carefully cut away the opening and trim right next to the line of stitching, being careful not to cut the pleated fab-

ric underneath. Turn your block to the wrong side and trim away the excess pleated fabric, leaving a ½" seam allowance. As you couch the rest of the block, place the yarn directly on the edge of any pleated sections that you come across.

Adding The Couching

A layer of batting is placed underneath the block so that you will quilt your block as you add the couching. The couching is completed exactly as it was for Dramatic Diamonds and Couched Seminole. You will analyze the lines to couch the shorter ones first, and then couch the longer lines (see page 60).

The quilting designs were traced onto the background with

a water soluble pen, then couched with Candlelight™ yarn. If your quilting design includes small curves, you may elect to bobbin-draw the yarn rather than couching. Since bobbin-drawing is worked free-motion, this eliminates having to pivot around the tiny curves. Look at the Carolina Lily block on page 64. All of the lines were worked by couching, except for the feather motifs in each of the four corners of the block. These were easier to add by bobbin-drawing, since I could easily slide around the tiny curves with the feed dogs dropped (refer to "Bobbin-Drawing"). The bobbin-drawing was worked through the layer of batting, just as the couching was, to give the work a quilted look. Just

STEP 6

STEP 7

STEP 8

STEP 9

STEP 6: The pleated fabric is positioned *underneath* each of the marked spaces and pinned in position. Straight stitching is done directly on the marked line.

STEP 7: The top fabric is carefully trimmed away to reveal the pleats.

STEP 8: Before couching, a layer of fusible batting is placed behind the background and lightly fused in place (avoiding the pleated areas). The batting is then trimmed away from each of the pleated sections and the batting fused firmly in place. This reduces the bulk underneath the pleated sections.

STEP 9: Quilt block ready for couching. (See completed Carolina Lily block on page 64.)

fuse the freezer paper to the batting and use a light box to trace your desired quilting design in position. The freezer paper will tear away from the batting when you have completed all of the bobbin-drawing.

You can see how it becomes automatic to combine the different techniques that you have learned. In fact, I hardly ever work exclusively with one particular technique. The more machine art techniques you know, the more effectively you will be able to evaluate a design to determine the ideal way to duplicate it in fabric and thread. Sometimes you will work part of your design using your appliqué foot and feed dogs, and the rest of it using free-motion techniques. With experience, you will easily be able to determine what to use when.

If you want to make several blocks of the same design, you can re-use your freezer paper for placement lines. Simply peel it away from your block after you have fused all the appliqué shapes in position, and then fuse it to the wrong side of the next block. This way, you only need to trace the design onto freezer paper one time, and all of the blocks will be identical. Have fun with this technique – create your own very unique sampler quilt!

A PRACTICE EXERCISE: PINWHEEL STAR

12-INCH BLOCK

SUPPLIES:
16" square of fabric
One ¼ yard piece of fabric for large star
Scraps for smaller pointed star and center circle (two colors)
½ yard Wonder-Under™
16" square batting (fusible or regular)
Couching yarns and threads (Metallic yarns and cords, trims, ribbons, etc; threads to match)
Sharpie™ pen
1 sheet template plastic
Optional: jewelry findings for embellishment

BELOW: Beautiful couched quilt blocks can be made from pieced and appliqué patterns found in quilt pattern books. This SUSAN'S WREATH block is based on an original pattern designed by Dolores Hinson and published in her book, *A Second Quilter's Companion* (Arco Publishing).

Let's examine this quilt pattern, Pinwheel Star, to determine how to prepare it for couched application. This pattern is one I designed for students learning to make couched quilt blocks, but it is similar in structure to many traditional blocks you will find illustrated in books. Provided is a scale drawing of the entire block, as well as the full-size pattern pieces. This is all of the information many pattern books provide, and it is enough to allow you to easily work with the block.

You will need to evaluate the pattern to determine the best way to enlarge it to full scale. Since Pinwheel Star is a 12" block, your first step will be to draw a 12" square on a piece of paper. In the scale drawing to the right you can see that Pinwheel Star is a symmetrical pattern radiating outward from the center. The points of the A shapes line up with the four corners of the square, so bisect your 12" square in both directions, from corner to corner, drawing lightly with a pencil.

Then trace Templates A, B and C onto a piece of template plastic and cut them out. Be sure to trace the placement guidelines on Templates A and C. You will use these templates to trace the pattern pieces in position on your full-size drawing, referring to the small scale drawing, top right, for placement.

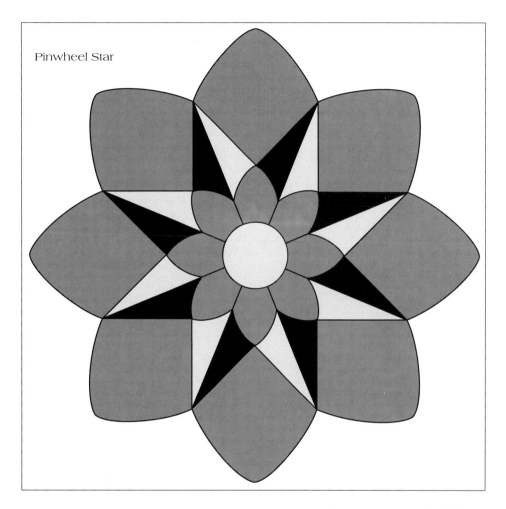

Pinwheel Star

First, trace the circle, Template C, right in the center of the square, using the guidelines on the template and the square for accurate placement. As noted above, the points of the A shapes line up with the four corners of the square, so your next step will be to place Template A with the bottom touching the circle and the point centered along one of the pencil lines on the square. Trace around this template, reposition it in another corner, and continue until all four corner A shapes have been drawn. Then draw four more A shapes, in between those already drawn, making certain the slides align and the bottoms touch the center circle. You will have a large star when you are done.

Now, place Template B in position, referring again to the scale drawing. You will see that the tip of the B shape's point touches the dip between the A

Shapes. This will be your reference point for positioning Template B. When you are enlarging patterns, these are the types of reference points that you should look for. Lay the longest edge of Template B along the long edge of Template A, with the tip just touching the dip between the A shapes – the reference point. Trace around Template B.

Referring back and forth to your scale drawing, trace the remaining Template B's. You will find that you need to flip Template B and trace it in reverse for half of the B shapes, in order to form the large points. When you

TEMPLATES FOR
PINWHEEL STAR

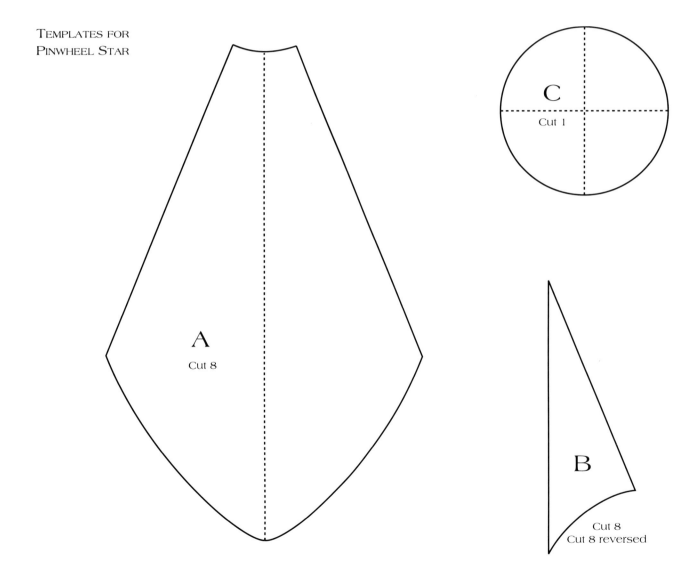

C
Cut 1

A
Cut 8

B
Cut 8
Cut 8 reversed

are finished, you will find that a smaller star has been formed automatically in the center!

When your full-size drawing is complete, trace it onto a piece of freezer paper and fuse it to the wrong side of a 16" square of background fabric. There is no need to prepare a mirror image tracing of Pinwheel Star because it is a symmetrical design. Use the templates to trace your patterns for the applique pieces onto the Wonder-Under™. Then select the fabric that you wish to use for the design and fuse the Wonder-Under™ to the wrong side of each fabric and cut the shapes out. All that is left is to place the background over the light box, use the placement lines that show through to position each appliqué shape, and then carefully fuse the appliqué pieces in position.

You are now ready for couching. Before I start couching a block, I often place a piece of batting underneath and pin it in place. (You may prefer to use fusible batting or fleece.) I couch any lines that might be added with a quilting stitch in the traditional pattern. You can mark such lines with a washable marker or with a pencil. Couch along all raw edges of your appliqué shape, using any combination of yarns and stitches that you wish. Refer back to the directions under "Couched Quilt Blocks."

You will end up with a beautiful block ready for finishing. Couched blocks are very attractive framed, used in wallhangings, made into pillows, or even sewn together with others to make a full-size quilt. However you display your block, you are sure to receive many compliments!

Plate 1
GUARDIAN OF THE
ENCHANTED DIAMONDS
This striking evening ensemble is exquisitely appliquéd with paisley-like flowers and ferns. Dramatic diamonds of various shades of blue lamé are connected with couched metallic yarns. Since dark values were carefully selected for all of the appliqué fabrics, the design, although it is fairly complex, does not appear too busy. This is an effective way to use elaborate appliqué to create a very dressy, sophisticated ensemble.

Detail shown on page 22.

Plate 2
CHERRY BLOSSOM
This very elegant silk jacquard suit is patterned with intricately appliquéd fruit motifs that are outline couched with red metallic yarn. Careful attention was given accentuating without overpowering the brilliant red jobot draping the right front. Cut-out sections of the unlined jacket reveal glimmers of the blue silk pantsuit underneath.

Detail shown on page 37.

Plate 3
SUNSET OVER DIAMONDHEAD
The rough textured boucle of this hand-painted silk jacket provides a striking contrast of texture for the padded silk charmeuse roses. The roses and the bodice of the evening gown are marbleized and embroidered with metallic threads. The jacket's softly draped closure, further emphasized by couching, subtly repeats the swirl effect of the sarong wrap skirt of hand-painted silk charmeuse. The metallic threads and yarns used for the applique, embroidery and couching help to balance the strong glimmer of the gold lamé diamonds at the cuffs and jacket hem.

Detail shown on page 36.

Plate 4
TIMELESS SHADOWS
This wool crepe jacket is embellished with dozens of brilliantly colored jacquard prints cut into triangles and couched with Ultramarine metallic yarn. Some of the triangles have been reverse appliquéd with pleated inserts to add an intriguing textural element. The multi-colored fabric used to construct the blouse provided the color scheme for the shaded flowers that have been machine embroidered on the shoulders.

Detail shown on page 58.

Plate 5
TROPICAL RENDEZVOUS
This wool jacket is embellished with tropical birds and flowers machine appliquéd in shiny polyester fabrics. The fine detailing of feathers and bud tendrils is worked in machine embroidery with rayon and variegated metallic threads.

Details shown on pages 20, 35.

Plate 6
TROPICAL RENDEZVOUS
(back view)
The pullover dress of black tissue faille has front and back inserts couched with metallic yarn. Birds and flowers flow from the shoulders and around the front and back in a V-shape to the bottom of the dress.

Plate 7
SPIRITS OF THE NIGHT WIND
The full-length collar and facings of this taupe Ultra-suede® coat are covered with hand cut polyester jacquard and Ultra-suede® squares that are couched with copper metallic yarn. The coat is appliquéd with swooping owls of richly textured polyester jacquards. The shading of the owl's details and feathers is worked in machine embroidery with a straight stitch, using dozens of shades of rayon thread. Hand-painted silk leaves are appliquéd, shaded, and connected with couched vines of metallic yarns. The facille skirt is appliquéd with silk leaves, shaded with embroidery.

Details shown on pages 17, 23, 61.

Plate 8
MONKEY ON MY BACK
This loose-fitting jacket has color blocked sections of silk noil and is lined in a richly patterned jacquard. It is appliquéd in abstract leaf designs that have been machine embroidered and textured with nubby yarns, metallic cords, and frayed silks. The over-size collar drapes in graceful folds down the center front. The back of the jacket features a mother monkey sitting atop an arched branch with her long arms tenderly encircling her baby. An Ultra-suede® belt was appliquéd to match this striking ensemble.

Details shown on pages 21, 91, 106.

Plate 9
AUTUMN GLORY
The collar of this tunic jacket of Jersey knit is covered with "Couched Seminole" squares of brilliantly colored jacquard prints. Partially decayed leaves, appliquéd in the same lustrous fabrics, appear sprayed in brilliant autumn colors. The tight curves and sharp points of the realistically shaped leaves are appliquéd free-motion, with the feed dogs dropped.

Plate 10
DIAMOND TRELLIS
This fitted jacket of emerald green, violet and tan silk noil is splashed with "Dramatic Diamonds" of shiny metallic lamés and brilliantly·colored jacquard prints. Elaborate couching with various metallic yarns along the diamonds adds color and texture. Trailing vines are bobbin drawn with emerald metallic yarns and clusters of jewel toned pansies are appli-quéd and realistically shaded with machine embroidery.

Details shown on pages 4, 6, 8.

Plate 11
SNAKES IN THE GARDEN
The front of this tissue faille blouse is appliquéd with slender strips of a snakeskin printed Ultra-suede® that have been cut apart in tiny squares and couched with metallic yarn. Stylized Ultra-suede® flowers sprout among the camouflaged serpents.

Detail shown on page 42.

Plate 12
THE BEAST WITHIN ME
This sublime silk suit is enveloped with tropical plants and majestic lions, all created with machine appliqué and embroidery. The narrow bands separating the jungle scenes are filled with tiny, ¼" couched Seminole squares of China silk. The lions are appliquéd in a single polyester jacquard fabric and all features and shading are machine embroidered with a straight stitch using dozens of shades of rayon threads. An Ultra-suede® hat was appliquéd to match the suit.

Details shown on pages 23, 50.

Plate 13
PATH BY THE FALLS
The free-flowing appliqué design of graceful ferns and tropical flowers blends perfectly with the loose design of the full, cape-like sleeves of the garment. Trailing vines are bobbin drawn with purple metallic yarn. The belt was constructed from the same metallic lamé used for the appliquéd flowers.

Detail shown on page 24.

Plate 14
MYNA RIOT
This loud, sporty outfit is a fitting trib-
ute to a very loud bird. The rayon/
polyester jacket features cut-out
squares that have red metallic fabric
inserted and couched with red
metallic yarn. Brilliantly colored satin
birds are free-motion appliquéd.

Details shown on pages 25, 106.

Plate 15
NEW YORK RED
Attention is drawn upward to the color-blocked bodice of this fitted silk poplin dress. Abstract shapes are appliquéd using a decorative computer stitch, and snips of ravely silks and metallic yarns are bonded in place and embellished with machine embroidery.

Detail shown on page 52.

Plate 16
LET'S GO TO MARKET
This is a Southwestern version of the infamous poodle skirt of the 50's! The lower skirt band has been couched and appliquéd in Seminole-like shapes of brightly colored hand-dyed cottons. The llama has been appliquéd and shaded with machine embroidery: his wooly pelt has shavings of mohair and wool that have been fused and embroidered in place. His market basket is grass cloth that has been machine embroidered in padded satin stitch in brightly colored bands of rayon thread. The ties around his neck are embroidered and tiny Guatemala dolls ride along for a guaranteed adventure wherever they travel. The bolero jacket has couched loops of nubby variegated yarn to resemble country hillsides, and the miniature country folk picking flowers, pulling wagons, riding horses, holding hands, etc. are all free-motion embroidered with padded satin stitches in brightly colored rayon threads.

Details shown on pages 28, 63, 86, 113.

Plate 17
UNTAMED SPLENDOR
This loose-fitting jacket of tissue faille is appliquéd with exotic silk roses. The shading of the flowers, vines and leaves is machine embroidered with a straight stitch using dozens of shades of yellow through red-orange rayon threads and many graduated shades of green. The wild snakeskin printed skirt is synthetic suede.

Details shown on pages 34, 104.

Plate 18
UNTAMED SPLENDOR (back view)
The jacket pattern was cut from plain
newsprint and machine basted together;
then the roses were designed directly on
the newsprint and carried across the
seam lines. This way, the appliqué design
flowed uninterrupted around the shoulders, down the sleeves, and across the
sides.

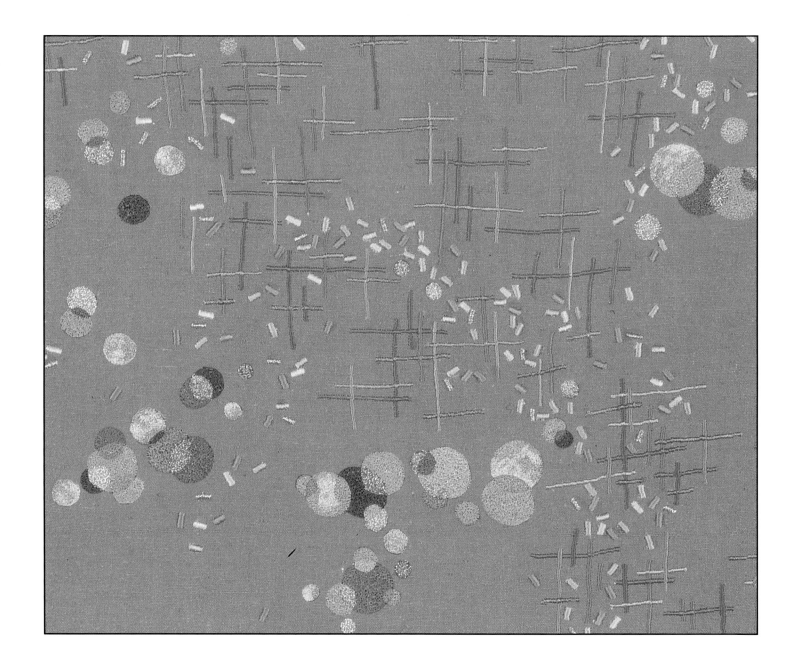

6. Machine Embroidery

You are going to *love* machine embroidery! You may not love it at first – it may take a while for you to feel comfortable working free-motion, and you may get discouraged and even disgusted with your early attempts – but, I promise that if you stick with it, learn to relax and play, you will soon experience the freedom machine embroidery gives you to create designs that would elude you were you using more conventional techniques. You will no longer be restricted in the direction you can move or the length or shape of the stitches you can use. A much wider variety of threads and materials will be

LEFT: With the feed dogs of the sewing machine dropped, the machine artist can travel in any direction, and a whole new world of creative stitchery opens up.

available to you and a much wider range of design possibilities will open up. In fact, I am convinced that absolutely anything that can be drawn or traced on paper can be reproduced using free-motion techniques.

I feel it is necessary to emphasize the many benefits gained by working free-motion because I realize that it is an intimidating technique to master. I won't ever forget the very first time that I dropped the feed dogs of my sewing machine to work free-motion. It was the oddest feeling, and I was overwhelmed by the decisions that were left open to me. Do I move forward, backward, sideways, diagonally, or perhaps in circles? Do I press my pedal fast or slow; do I move my hoop fast or slow? How do I coordinate the

movements? Do I take short stitches, long stitches, stacked stitches, straight or zigzag stitches? Of course, the beauty of machine embroidery is exactly what I just described. The freedom to make these decisions and combine them to create infinite possibilities!

In learning machine embroidery, it is especially important to practice. As with all eye/hand coordination skills, it will take a while before your movements become consistent and you gain the confidence and proficiency to create on fabric the stitches you envision in your mind. And as with all other skills that take patience to master, once you become competent, the technique becomes very easy. You will wonder how it could ever have felt awkward!

The ideal way to reach this

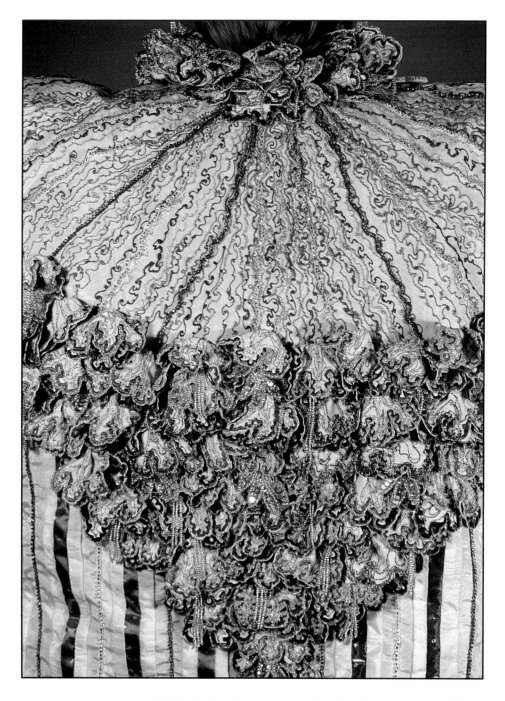

ABOVE: Marilyn Boysen was inspired by the wonderful textures of lichen to embroider these gorgeous free-flowing Ultra-suede® appliqués for her garment, "Icy Winter Storm."

level of expertise is to not concern yourself with what your early attempts look like. It doesn't matter what they look like! What is important is that you learn to associate a particular stitch formation with a particular movement. In other words, say to yourself, "Okay, I created this stitch – I either love it or hate it, but exactly how did I create it?" File this information in your mind so that in the future you can either reproduce or avoid that same effect. What may seem inappropriate in one situation may be desirable for another. That's why it is important to document, either in your mind or on paper, the progress of your practice stitches.

I encourage my students to save the scraps of fabric with their practice stitches, and with a fine tipped permanent marker write the width and tension settings, type of thread, and any other pertinent information directly on the fabric. In fact, I encourage any machine artist, at any level of expertise, to save such scraps of fabric with important practice stitches. I keep a plastic bin full of such scraps, which I add to whenever I experiment with new stitch combinations or techniques. You think that you will remember what you have learned through trial and error, but often you forget. These fabric and thread mementos will help you remember details and save you time when you are experimenting in the future.

Many people are reluctant to retain these experimental scraps because they are usually not very pretty to look at. These same people are the ones who become disgusted with the way some of their stitches look and give up! It is okay to become disgusted – just don't let your frustrations triumph! Learn to play just for the joy of playing, and not worry about creating beautiful designs until you feel comfortable stitching free-motion. The designs will evolve with a life of their own once you feel confident to take risks.

I am afraid all this introductory material may make some students nervous about attempting machine embroidery, when in fact what I want to emphasize is that machine embroidery is easy and fun, once you decide it is easy and fun. I have seen many students want so badly to create beautifully shaded flowers with their first attempts, that they become extremely frustrated when their first stitches don't meet their expectations. Other students jump right in and doodle and draw, scribble and shade with great enthusiasm, because they are truly enjoying themselves and don't expect controlled stitches right away. Machine embroidery, more than any other skill I have learned or

taught, is much a matter of your frame of mind. So think positive!

Let's start some of those practice scraps and have fun and learn!

MACHINE SET-UP

1. Thread your machine with machine embroidery thread and fill your bobbin with a contrasting thread. The bobbin may be either cotton machine embroidery thread or regular dressmaker's thread. Don't ever put rayon embroidery thread in your bobbin.
2. Put in a size 70/10 needle.
3. Slightly lower your top tension setting.
4. Remove your sewing foot.
5. Lower feed dogs, or use a cover plate.

For the first exercise, you will need a machine embroidery hoop and a 12" square of muslin. Be sure to use a machine embroidery hoop, as the ones manufactured especially for machine embroidery are much stronger than craft hoops and tighten much more securely. Also, they will have a shallow groove cut from the top to allow the hoop to slip underneath the needle. When filling the hoop, align these grooves so that the needle can pass over the edge. However, test your own machine because on some models the needle will lift high

enough so that this will not be necessary.

Machine embroidery hoops come in a variety of sizes; you will probably want to have several different sizes on hand. The smaller hoops stretch the fabric the most taut and are perfect for small designs. The larger hoops give you more space to embroider larger designs, but do not hold the fabric as tightly. You may find it necessary to re-tighten your fabric if it slips as you work. Also, the larger hoops will hit the arm of your machine as you advance. When this happens, simply swivel the hoop so that the bulk of the hoop is to your left.

I have found it very helpful to wrap the inner ring of my hoops with strips of pantyhose to prevent the fabric from slipping once it has been tightened. Also, this extra cushion between the hoop and the fabric helps take up any slack if the fabric has uneven relief (thickness). When you embroider a portion of your design and then move the hoop to another area, the embroidered section positioned between the hoops will cause unequal tension. The pantyhose will fill in any gaps between the embroidered area and the hoop.

Just cut a length of pantyhose in a 2" strip, and wind it around the inner ring. When you return to the start, cut the pantyhose and secure it with tiny stitches at the inside of the ring. Of course, some people may be adverse to using pantyhose as it does tend to de-glamorize a hoop! You can also use bias tape, or any soft fabric that does not ravel, but I have found that pantyhose works perfectly!

To fill your hoop: Place the outer ring (it will have the screw at the top) on a table and center your fabric over it. I like to stand for this step as I have better leverage. Place the inner ring on top and firmly press it to the inside. The fabric should snap it tautly. Check for any puckers and straighten the fabric if necessary. The fabric should be stretched drum tight; this is important. If the fabric is too loose, your machine will not be able to form a correct stitch. As

you stretch your fabric, be careful that you do not pull it off-grain. After it is securely stretched, push the inner ring about 1/8" inside the outer ring. This way, the inner ring will be pressed firmly against the throat plate when you embroider. Now tighten the screw so that the fabric will not slip.

EMBROIDERING SHAPES

Lift your presser bar and place your hoop beneath the needle. Take one stitch to bring the bobbin thread to the top; hold both threads to the back behind the needle to start. With your machine set at straight stitch, run the sewing machine and practice doodling. You will not pivot; that is, you will not

turn the hoop. Instead, when you want to change direction, slide the hoop to either side or forward and backward. If you catch yourself rotating your hoop – stop! The exercise shown in Fig. 6-1 will help you to become accustomed to the correct way to follow lines free-motion.

Practice guiding your hoop to stitch these squares. Do not trace the design; rather, repeat to yourself as you move the hoop along: "Left, backward, right, forward, left, backward, right, forward." These are the directions you will be sliding the hoop to create the right angles; soon it will begin to feel very comfortable and automatic.

You will notice that the faster you move your hoop, the longer

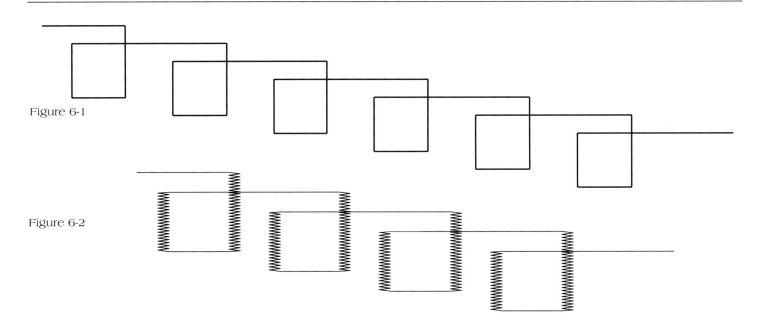

Figure 6-1

Figure 6-2

your stitches will be. You are in control of stitch length, so if you desire shorter stitches, move your hoop more slowly or move your foot pedal faster. Part of the skill for machine embroidery comes from gaining control over these variables. You want to embroider at an even pace while being able to control stitch length and direction. If you have a "sew slow" button on your sewing machine, you may find it helpful to engage it and sew with your foot pressed firmly on the foot pedal. This allows you to become accustomed to a set speed; you can develop rhythm while being confident that your machine won't "take off."

Once you feel comfortable drawing squares with a straight stitch, switch your machine to a medium-wide zigzag and practice the same outline. You will notice that as you sew sideways, your stitch formation will be very thin, but as you pull the fabric either forward or backward, the stitches turn into satin stitches (Fig. 6-2).

To form a pretty satin stitch, you will need to pull the fabric forward (or backward) *very* slowly while pressing the foot pedal hard. It will take practice before your stitch length becomes consistent, so don't become discouraged if your first row of satin stitches looks lumpy and crooked! When you are pulling your fabric sideways to form the side stitch, you will want to slow down your machine speed. These are the two basic stitches for machine embroidery: the satin stitch and the side stitch. Notice that you achieve these stitches not by using different stitch settings, but by pulling the fabric in different directions.

On your scrap piece of muslin, practice different random designs: circles, angles, crosshatch lines, giant zigzags – whatever you feel like doodling! Try the same type of lines first with a straight stitch, then switch to zigzag, and practice controlling your speed. Experiment wildly, as shown in Fig. 6-3. Play, play, play!

Figure 6-3

You should be having lots of fun by now. You are no doubt becoming mildly impatient to create something beautiful. Do you know that you are already an accomplished machine embroiderer? All we need to do to turn those doodles into works of art, is to substitute many different thread colors as you practice. Really! Try the following exercise, using a firmly woven fabric for your background and at least half-a-dozen luscious colors of rayon machine embroidery thread and at least one metallic thread.

Select a color, and with your machine set at straight stitch, start making tiny, wandering curves. Floor your foot pedal, and make these curved lines closer and closer together until you are able to fill in concentrated areas. Meander across your fabric, continuing in loops and curves and varying the density of your stitches. Your fabric will resemble Fig. 6-4.

Enough of that – it's time to change thread color. Repeat the same meandering stitches with your new color; blending next to the first stitches and then traveling off in other tangents. Again, fill in concentrated areas with tight stitches, and then take longer, meandering stitches in other areas to keep the design light and airy. It looks better, doesn't it? You will notice your new designs emerge with each color you add. When you work with metallic threads, it will be necessary to slow down your speed, as these metallic threads are much more fragile than the rayon. If your needle goes down in an embroidered area, metallic thread will fray and then break. I am aware of the extra care necessary to embroider with metallics and don't mind slowing down, and re-threading when necessary. I think these beautiful threads are well worth the extra precautions necessary to add their sparkle to my designs!

When you are happy with your design, you might want to add some confetti-like padded satin stitches for a pretty contrast in texture. Set your machine to medium-wide zigzag and select a new color. Now, pull your fabric forward to form a row of satin stitches about ¼" long. Next, pull your fabric backward with the machine still running, and then slowly forward again. As you embroider over your satin stitched row, it will appear padded. To tie off, set your machine on straight stitch and take three or four stitches at the edge of your satin stitches (unless, of course, you have that handy "tie-off" button). Lift your presser bar, travel to another area, pivot your hoop to change the direction of your next row, then lower the presser bar to continue. It is not necessary to cut the bobbin thread as you embroider. When you are finished, you can trim all the loose bobbin threads on the back (see photo on page 79, bottom left). When you know you are finished (sometimes this point is hard to recognize!), steam block your fabric while it is still stretched in the hoop. Hold your iron over the back of the hoop and let the steam from your iron push into the fabric. Release your fabric from the hoop, and place it on an ironing board, face down, to finish pressing. Your embroidery is complete!

Figure 6-4

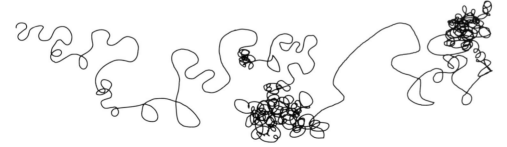

The photo at bottom right illustrates more design possibilities with the padded satin stitch. Vary the width of your bars, keep them perpendicular to each other, and have fun integrating new colors.

SHADING

Now let's try something a little different. We are going to practice *shading* which is simply blending colors next to each other by feathering the stitches. Select three or four colors of thread that you think will look attractive blended together; you may want to include a metallic. Starting with a strong color, set your machine to medium-wide zigzag and side stitch a line about 2" long. Now fill in a small area along the line, constantly moving your hoop from side to side and keeping both edges very jagged (Fig. 6-5A).

Thread the machine with your next color, and continue where you left off, blending your new color into the first stitches at the jagged edge. For the most subtle blending, you will want the edge where the stitches feather into each other very jagged (Fig. 6-5B). Don't be con-

BOTTOM LEFT: Meandering stitches of loops and curves, all worked in a straight stitch using several different colors of rayon and metallic threads; some areas are concentrated with stitches while others are kept light and airy. The confetti-like designs are padded satin stitches.

BOTTOM RIGHT: The padded satin stitch: vary the width of your bars and keep them perpendicular to each other.

Figure 6-5

A

This   Not This!

B

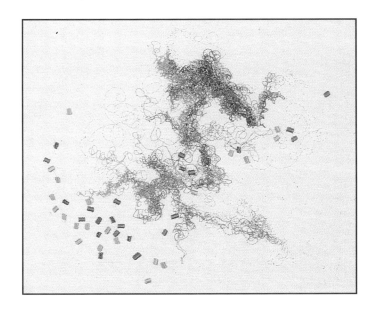

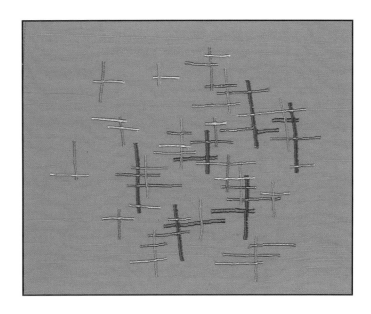

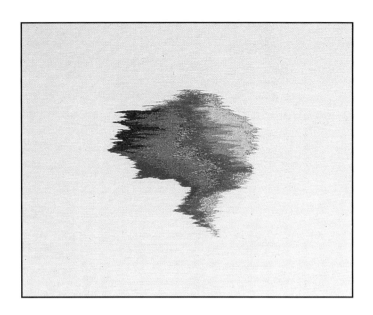

cerned about creating a particular design; just enjoy practicing blending your colors together. You will be pleasantly surprised when a pretty pattern materializes! (see photo above, left) You can shade using a straight stitch also, as illustrated in the photo above, right. The principle is exactly the same: feather your stitches where two colors blend. To create the fan shaped design in Fig. 6-6, stitch in and out, radi-

ating from the center. Slightly curve your lines so that they won't appear stiff; and be sure that after you travel out from the center, you stitch back on top of the same stitches to return to the center.

Now let's try shading in a more controlled area. Trace a circle, approximately 2" in diameter, on a 10" square of muslin and stretch it in your hoop.

Select three or four colors of thread to shade with and thread your machine with the first one. Set your machine to medium-wide zigzag and carefully satin stitch the left edge of your circle (Fig. 6-7A). Remember, do not *turn* your hoop, but rather, *slide* it as you travel from top to bottom. Now travel to the top of the circle again, moving back and forth sideways to create your

Figure 6-6

Figure 6-7

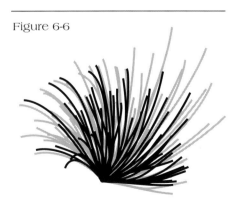

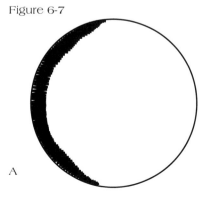

A

B

OPPOSITE PAGE
TOP, LEFT: Shading with a side stitch.

TOP RIGHT: Shading with a straight stitch.

RIGHT: Draw overlapping circles and fill them in with shaded side stitches using graduated colors of rayon and variegated metallic threads.

feathered edge (Fig. 6-7B). Moving constantly with a side stitch, fill in the middle ground, keeping the right edge jagged (Fig.6-7C).

Change to your next color, and blend it next to the first one at the left edge. Continue as before, adding new colors as desired, always keeping that right edge jagged for blending in your next colors (Fig. 6-7D). Don't be concerned with keeping the colors divided equally. When you approach the right side of your circle, carefully satin stitch the edge (Fig.6-7E), and then fill the inside with side stitches.

You can have a lot of fun and create very pretty designs by drawing several different size circles and overlapping some of them (see photo above). Color the overlapping areas in contrasting colors of thread. Experiment with variegated rayon and metallic threads, and if circles are very small, you can fill them in with straight stitches.

I think it is important to practice on geometric and abstract shapes until you feel at ease working free-motion. You must feel comfortable working free-motion, have good control over stitch length and direction, and be able to easily flip back and forth from straight stitch to zigzag and adjust stitch widths, etc. before you attempt to create a shaded flower! My motto: patience for practice equals fewer frustrations!

Figure 6-7

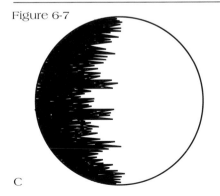

C

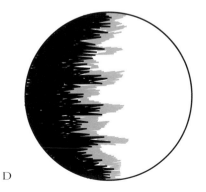

D

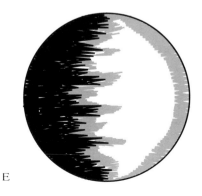

E

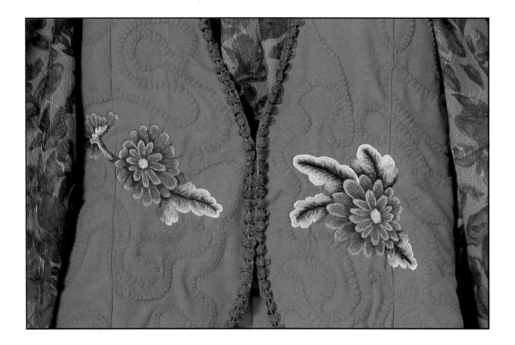

EMBROIDERING REALISTIC IMAGES

Now that you have completed all of the preceding exercises, by golly, you are ready to embroider anything! Let me illustrate how simple it is to incorporate what you have learned to embroider realistic images. For example, you can recognize that the flower petals shown in the photo above were shaded exactly the same way we did the circles. The main difference is that you will want to be more selective in the colors you choose for shading. You will want to choose graduated values, at least four shades, from light to medium to dark, of a single color. The closer the shades are in value, the more realistically shaded your flower will be.

Also, the direction in which you embroider is very important. For the smoothest result, curves should be satin stitched and points should be side stitched. Therefore, you always want the needle to swing horizontally into a point or curve. Think of your needle as forming a horizontal line as it swings back and forth. You want the needle to swing into any curves or sharp points of your designs. When you embroider each petal of your flower, you are going to line the needle swing into the curved tip. (Note the direction of the arrows in Fig. 6-8A.) You can see that it will be necessary to pivot before starting each petal separately, to line your needle up correctly.

Starting at the base of your petal with the darkest shade, embroider the bottom portion, feathering your stitches to blend the next shade (Fig.6-8B). Change to the next darkest shade and continue toward the tip in the same manner you embroidered the circle. When you approach the tip, satin stitch the edge with the darkest shade, and then fill it in (Fig. 6-8C).

When you are embroidering a flower, you can embroider the darkest shade in all of the petals at the same time; just be sure to

Figure 6-8

A   B   C

line your needle up for each separate petal. Trace the floral designs on pages 84 and 85 and shade them with your favorite colors! Be sure to line your needle up with the arrows.

As your embroidery skills improve, you may wish to practice without using a hoop. You will need to back your fabric with fusible stabilizer or freezer paper to keep it from puckering, and use a darning foot.

Getting used to the darning foot will take a little practice because your vision will be somewhat obstructed. After a while you will become accustomed to peering through the small opening of your darning foot, and the foot does have the advantage of protecting your fingers from the needle. I don't like the restrictions that the machine hoop imposes. As you have

probably noticed, I don't like any preparation or step that I consider tedious or unnecessary. I, therefore, work almost exclusively with my darning foot. This, however, is a personal choice, and it is important to be aware of your options. Many people prefer to use the machine embroidery hoop, and I find it necessary to use it for some designs.

TOP LEFT: 3-dimensional realism is achieved by thread painting in graduated shades. Notice how the iris petal subtly changes from blue to yellow toward the center.

TOP RIGHT: This basket of fruit, embroidered by Beverly Burton, illustrates how beautifully detailed images can be created with thread. Note the combination of the green threads, and how dramatically the black thread outlines the basket and separates each fruit and leaf.

Figure 6-9

Figure 6-10

MACHINE EMBROIDERY TIPS

- To start: Take one stitch, tug on your top thread to bring the bobbin thread to the surface, then take several tight stitches to lock it. Once started, you can lift your presser bar to "travel" (move from one area to another without stitching), then lower your presser bar and start again by taking several tight stitches. Every once in a while you will want to "clean up" the underside by trimming loose threads.
- To end: Always stop a line of straight stitches by stitching in place three or four times. If you are zigzagging, you will need to switch to straight stitch to tie off in the same manner. Or you can, of course, use your tie-off button if you have one!
- Treat metallic thread with kid gloves! Stitch slowly, do not stitch into previous stitches and take longer stitches, than usual. You will be rewarded for your precautions: metallics are gorgeous!
- If you are interrupted during a project, write the thread types and numbers on the back of your freezer paper for future reference.
- You can embroider your design on organza stretched in a hoop and later appliqué it to your garment or desired background. After you have finished the embroidery, remove it from the hoop and back it with Wonder-Under™. Trim right next to the stitching, remove the paper backing, position your design, and fuse it in place. Use a darning foot to appliqué the edges with matching thread, either with straight stitches or a tiny zigzag.

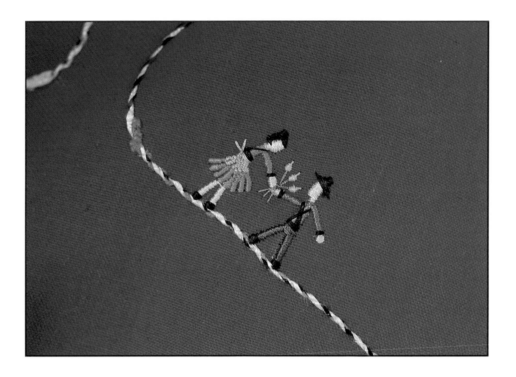

FOLK ART EMBROIDERY

The charming folk art people shown are embroidered with a technique called padded satin stitch. By slowly moving the fabric forward and then backward, the satin stitches that are formed can be built up until they appear very padded. The images are simple stick figures that are characterized by bright and bold colors of thread.

It is important to distinguish the different widths of the zigzag settings on your machine. In the pattern instructions in this book, the following settings are used:

small: 2.0 mm
medium small: 2.5 mm
medium: 3.0 mm
medium wide: 3.5 mm
wide: 4.0 mm

Even if your settings aren't marked by these particular numbers (which are millimeters), use the numbers as a guide to determine descriptions for your machine's settings.

Stretch a piece of bleached muslin in your hoop and thread a flesh colored rayon thread. Refer to the enlarged drawing (Fig. 6-11, page 89) for a guide as you stitch, but you will actu-

TOP & BOTTOM: Embroider tiny folk art people onto a special garment along couched hills of nubby yarn.

ally trace only the smaller stick figures in Figs. 6-12A and 6-12B, page 89, on your fabric.

Starting at the bottom of the girl's legs, stitch two rows of padded satin stitches, ending at the hem of her dress. You will be moving slowly up and down until you are satisfied with the padded appearance of the stitches, referring to the illustration for the stitch widths for each detail.

Change to a brightly colored thread for the girl's dress. Starting at the edge of one corner, stitch the bottom portion of hem in the wider stitch widths for each detail. Pivot slightly, so the dress will fan out, and stitch the next row in the same manner. Continue across, creating the bottom portion of her dress (Fig. 6-13, page 89).

In the same color, continue toward the waistband in the same manner, stitching in padded satin stitch rows, in the reduced width setting.

TOP: This charming scene was designed by Beverly Burton, who has admitted that she has become addicted to creating new settings for her folk art people! Walking the dog, sharing apples, and picking flowers are all merry activities these little folk like to engage in.

BOTTOM: These little folk will fit anywhere! In "Trailing Stars" they are embroidered inside couched diamonds.

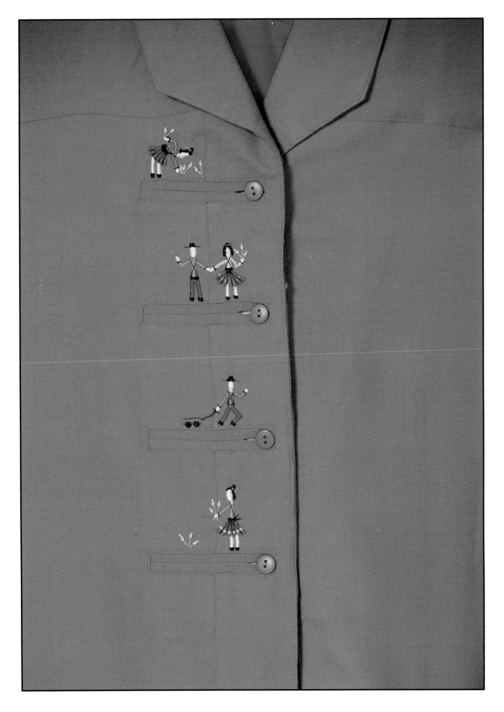

ABOVE: Embroider folk art people down the front of a shirt.

For her waist and bodice, stitch one wide row of padded satin stitches toward her neck. The skirt of the dress and the bodice will blend, and she will start to look like a real person at this point! Next, create her arms from the shoulder line in the narrower width setting. When her arm changes direction at the joint, pivot to keep stitching in a forward direction.

Change back to flesh and add tiny hands at the end of her arms. Also add the neck at the top of her trunk. For her face, refer to the face detail in Fig. 6-12C. At the top of her neck, you will embroider at the lower setting, padding the stitches back and forth, and then skip to the wider setting and continue padding your stitches. After the thread has built up equally, continue at the top, using the smaller setting again. After a little practice, you will be able to embroider very round faces!

Our little lady will really come to life next as we add the black embroidered details. First, add shoes at the bottom of her legs (patent leather of course!). For her hair, stitch a narrow row of black on each side of her face, with your needle swing barely biting into the edge of the flesh color. At the top, switch to the wide width to add her bangs; then switch to the narrow width again if you wish to add a tiny bun on top of her head. By

Figure 6-11

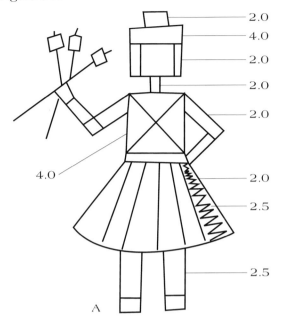

2.0
4.0
2.0

2.0

2.0

4.0

2.0
2.5

2.5

A

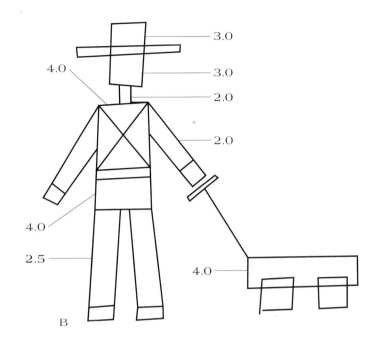

3.0

3.0

4.0

2.0

2.0

4.0

2.5

4.0

B

Figure 6-12

A

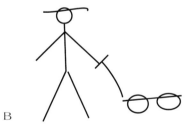

B

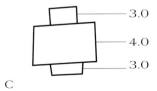

3.0

4.0

3.0

C

face detail

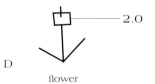

2.0

D

flower

Figure 6-13

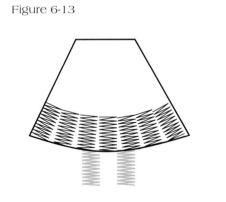

now you are discovering the secret of this magical padded satin stitch – the illusions are created by utilizing many colors of brightly contrasting thread.

You can add more details. If you want to add a band of contrasting color across her pleated hem, simply "walk" across the dress in the same width zigzag, just above her hemline. Take two or three zigzags in place at each pleat, pivot to line up for the next pleat, and continue across.

The rest of the details are worked using a straight stitch, and you will control your stitches by moving the flywheel by hand. With black, start at the waist and take a single stitch right at the division of one pleat. Stitch in place several times to lock your

thread, then carefully pull the thread to the bottom (lift the presser bar if you feel resistance). Bring the flywheel forward to take a stitch right between the pleats from the top to the bottom of the skirt. Then travel back to the waistband and take another locking stitch. Continue across, dividing each pleat with one long stitch. For her straps, at the edge of one side of her waist, take a locking stitch. Then carefully pull the thread to the shoulder at the opposite end, and then back to the starting point. Still at the waist, travel across to the opposite side, and then cross the first long stitch to the opposite shoulder, and back to the waist again. Then take four or five long stitches, always turning your fly-

wheel by hand, to embroider the black waistband.

The stems of the flowers are worked in long straight stitches also. Simply direct your stitches exactly where you want them, turning the flywheel by hand. For the tiny flowers, take one short straight stitch for the stamen, then switch to narrow zigzag (Fig. 6-12D, page 89) and stitch in place for 10 to 12 complete zigzags. The stitches will build up until they bow over, and they will appear to be rounded. This is sometimes referred to as *French knots by machine*. By now, you can tell at a glance exactly how our little man was created. For the wagon, turn your fabric sideways so that the row of padded satin stitches will be parallel to the ground. The wheels, more circles, are stitched exactly like the face detail and in the same width settings. The handle is a long straight stitch, topped by a shorter perpendicular straight stitch at the man's hand. His hat brim, also, is a long straight stitch.

These little folk can be very addictive, and you will find yourself designing others: flying kites, chasing dogs, playing ball, etc. And if you have a computer sewing machine, try adding other tiny details with your built-in embroidery stitches.

BONDED APPLIQUÉ

I believe you will love this enjoyable technique. If you are a diehard texture lover like I am, this may be one of your favorites, for it allows you to use lovely frayed fabrics, loopy threads, hairy yarns, frazzled laces and even leftover serged thread chains and turn them into beautiful collages. I keep a box nearby and toss in all the scraps of fabric and thread that are too beautiful to throw away. When I am ready for some exciting experimentation, I bond them to background fabrics and embroider them in place. What fun!

To get started, first collect anything fuzzy, furry, hairy, frazzled or stringy that you personally think is beautiful. Metallic threads, fabrics, and yarns are wonderful. So are worn out lace scraps, crocheted edgings, decorative serger threads, and Balger™ filament threads. All kinds of fabrics are great too; the more easily they fray, the better.

BELOW: Bonded appliqués of nubby yarns, metallic cords, frayed silk, and Balger™ filament threads will liven up simple and bold abstract shapes.

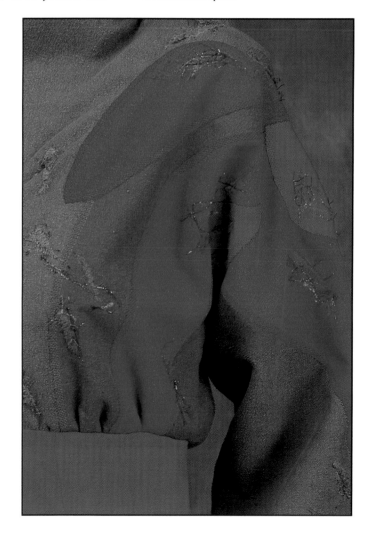

Now select a background fabric for practice, something firmly woven, preferably a natural fiber. You will also need a piece of Wonder-Under™ and a piece of transparent fabric (such as netting, organza, gauze, etc.). Both pieces should be the same size as your background. You will also need a Teflon™ Appliqué Press Sheet™ and some deli paper.

FUSING THE APPLIQUÉ PIECES

1. Separate the Wonder-Under™ from its paper backing. Reserve the backing and place the webbing on top of your background fabric.
2. Now for the fun! Shave slivers of fabric; ravel the edges and pull apart the weave and arrange over the fusible webbing. Sprinkle threads, yarns, torn bits of lace and ribbon on top. Curl long strands of metallic and rayon threads, snip apart fabric fur, arrange slivers of Ultra-suede®. You are limited only by your own imagination! The important thing is to experiment and have fun!
3. Place your transparent fabric on top of this textile collage, then place the teflon sheet on top. Press until the collage is bonded to the background; then remove the teflon sheet.
4. Use the backing of your Wonder-Under™ to remove the residue webbing by laying it

on top of the collage and pressing with your iron over the entire surface. When you peel away the paper, most of the residue webbing will adhere to the paper.
5. Place a sheet of deli paper on top and continue pressing with your iron through the paper until all the residue webbing has been removed. It will stick to the deli paper.

Note: Now you are ready to give your collage more permanency by embroidering the bits of fabric and threads in place. At the same time, you can add additional embellishment with the embroidery threads you select.

ADDING EMBROIDERY

1. Drop the feed dogs of your sewing machine, or cover them with a cover plate. Put a darning foot on and lower top tension slightly.
2. Cut a piece of freezer paper the same size as the background fabric. Using a hot, dry iron, fuse it to the wrong side of your collage. Press with the waxy side of the freezer paper against the wrong side of the collage.
3. Now is the time to carefully choose the thread you will use to secure and embellish your collage. If you want the thread to be inconspicuous, choose a color to match your background fabric or use

transparent nylon thread. If you want your thread to embellish your design, choose rayon machine embroidery threads in colors that complement and accent the colors of your collage. For special sparkle, you can use metallic threads. Again, experimentation is the key!
4. Stitching can be done with a straight stitch or a zigzag setting, (or a combination of both!). Be sure to stitch over any loose pieces of your collage that are not securely fused in place. The machine embroidery can be very controlled, with definite shapes, or it can be very random and loose, with spontaneous loops and crosshatching.
5. After the stitching is complete, carefully peel the freezer paper from the back, using tweezers. Steam press from the wrong side, smoothing out any puckers as you press. Your collage is now complete – and washable! Be proud, you definitely have a very unique design!

Note: Step 2 may be omitted if the background fabric is firm enough to not pucker when being embroidered. Simply test a small piece first without using the freezer paper and check for stretching or puckering – or if you prefer, use a hoop!

After completing your first

sample piece, you will have all kinds of new ideas for your next. Every project will be very different because of the infinite number of fabrics, threads, and other materials available for bonding. You will look at leftover scraps in a whole new light and be reluctant to throw anything away; even the multicolored fluff collected in the lint trap of your dryer will look pretty interesting! The best part is that this technique produces a very washable embellishment, for the more the fused bits become frayed and frazzled, the better they look.

The following samplers illustrate some variables you might want to experiment with.

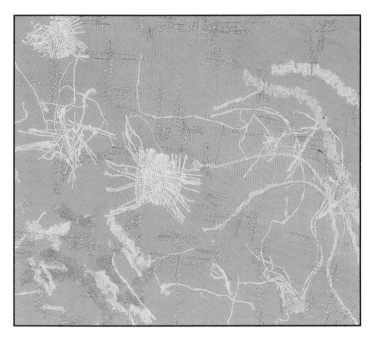

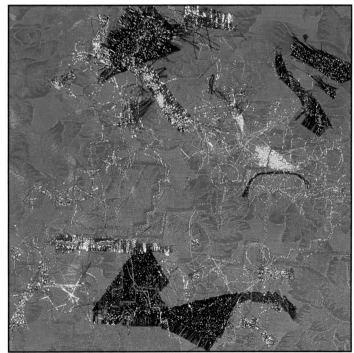

ABOVE: The background is brushed denim and the overlay is bridal netting, which has been layered in areas to create shadowing. Bonded materials are very loosely woven silks and linens and a single, variegated metallic thread is embroidered in crosshatched straight stitches.

ABOVE: The background is polyester jacquard and no overlay has been used. Very ravely metallic fabrics and threads have been bonded and embroidered in place with zigzag stitches of ultra-marine rayon and variegated metallic threads.

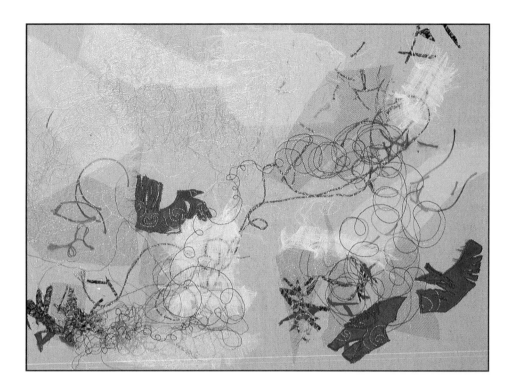

LEFT: The background is silk poplin, and various snips of overlay materials have been placed at random: plain and tinted bridal netting, organza, batiste, Twinkle cloth. Bonded materials include pearl rayon (a decorative serger yarn), metallic yarn, rayon embroidery thread pulled from the spool and piled in bunches, scraps of satin and polyester jacquard scraps cut in narrow shreds. All have been embroidered in place with straight stitched open loops of metallic and rayon embroidery thread.

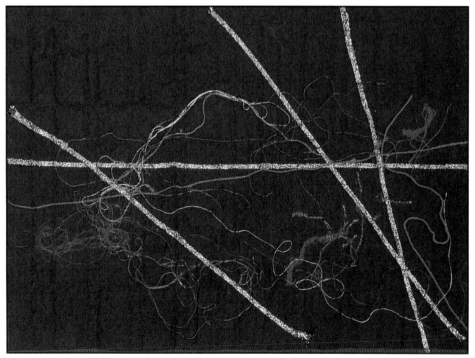

LEFT: The background is navy rayon and has polyester fleece bonded to the wrong side. No overlay has been used. Bonding materials are all threads and ribbons, this time several strands of a wide metallic yarn have been laid at straight angles to create a more structured design. The embroidery is navy rayon stitched in parallel cross rows that create softly quilted depressions through the layer of fleece.

RIGHT: For a more subtle effect, bonding materials and embroidery thread were chosen to match the background fabric in color and value. The background has also been layered with batting and quilted as it was embroidered.

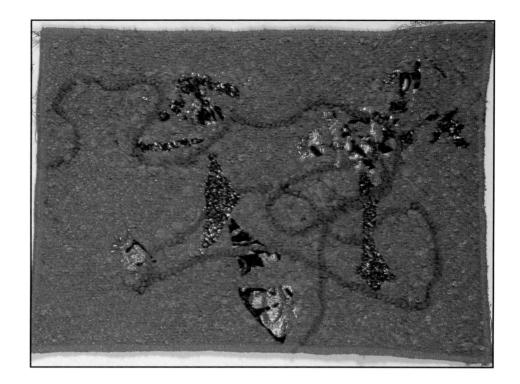

RIGHT: This may appear as abstract as the other samples, but it is actually representational of a small bouquet of flowers. Instead of scattered randomly, the bonded materials (metallic yarns, threads and fabrics) were clustered by color and shape to represent stems, leaves and blossoms. The background is linen, and smaller scraps of organza and colored netting are trapped beneath a full layer of sheer organza. The embroidery is straight stitched with silver metallic thread in broad loops and swirls.

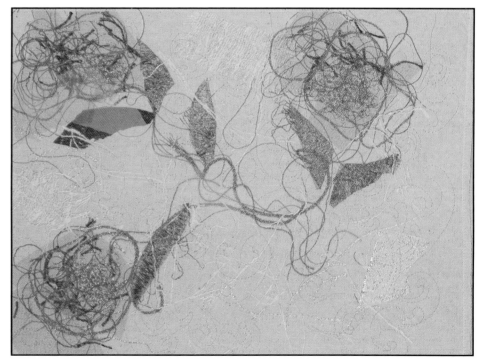

I made the little tote pictured below to carry my class handouts. The combination of textural silks and metallics creates an aura that instantly transports me to the lush setting of my friend's tropical greenhouse, where the photo which inspired the tote was taken. Fuchsia-colored netting which matches the cotton background is effectively used as an overlay to blend the exotic mixture of colors and textures.

BELOW: Another abstract picture, this block is a representation of a potted plant in a tropical greenhouse.

BOBBIN DRAWING

This technique will allow you to draw with heavy threads that do not fit through the eye of your sewing machine needle. It is a technique that is largely overlooked because it entails making adjustments to the bobbin tension. I have already stressed the importance of understanding how the tension of your bobbin functions, and knowing how to adjust it to achieve the effect you want. Now you are going to discover just how exciting it can be to manipulate the bobbin in order to utilize those intriguing threads for machine art.

Choose a specialty thread (from now on, referred to as *bobbin yarn*) from the large selection available: metallic yarn, pearl cotton, ribbon floss, embroidery floss, etc. I maintain the same attitude toward thread as I do fabric: I will try anything. If it doesn't work, fine, I know I can use it for another technique. Remember, experiment.

You will want to select a rayon, metallic or regular thread which matches the bobbin yarn for the top spindle of your sewing machine.

You can fill your bobbin with the bobbin yarn by hand (very tedious, you will quickly discover), or you can wind it using your sewing machine. Do not thread the yarn through the normal bobbin winding path; rather, set the bobbin its usual spindle but set the bobbin yarn's spool on the end of a pencil held in your right hand. Pinch the yarn between the fingers of your left hand to give it a little tension as you press the foot pedal to engage the winding mechanism. Fill the bobbin, but not so full that the yarn extends beyond its edge.

Now you will need to adjust your bobbin tension. You want the bobbin yarn to feed with the same amount of tension as regular or embroidery thread. Since the bobbin yarn is much heavier, it will be necessary to release the tension slightly. Using the tiny screwdriver, turn the tension screw counterclockwise, loosening the tension just until the bobbin yarn feeds smoothly when you pull on it. You will soon become accustomed to what it should feel like. Remember, if you loosen it too much, it is a simple matter to tighten it up a bit – don't be intimidated! Some people prefer to purchase a spare bobbin case for the sole purpose of bobbin drawing, which is fine, but is certainly not necessary.

Machines that do not have a separate bobbin case will have a tension bypass instead. If your machine doesn't have a separate bobbin case, check your manual for instructions on bypassing the tension of your bobbin and thread your bobbin accordingly.

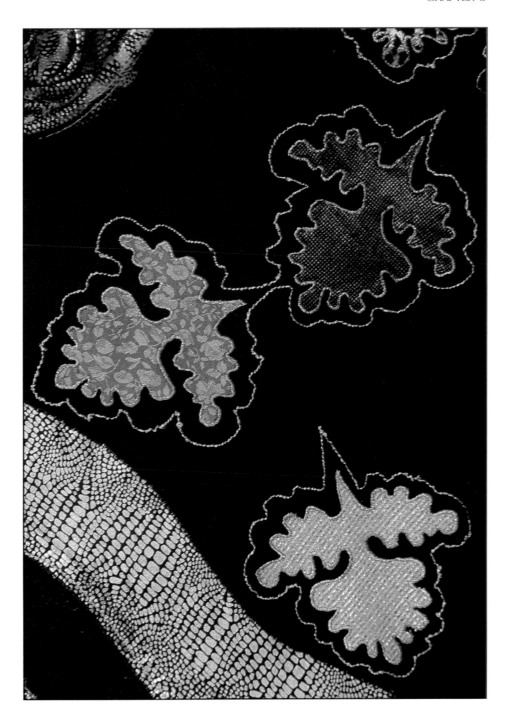

ABOVE: The bobbin drawing with metallic gold yarn adds movement to these stylized leaves.

Since your decorative thread is in the bobbin, you will be sewing upside down. (Did I forget to mention that?) Actually, sewing upside down has certain advantages. I like to trace my design onto freezer paper. Then, after fusing it to the wrong side of my background fabric, I have an outline to follow without having to mark my fabric. You can use a machine embroidery hoop or use the freezer paper method in conjunction with your darning foot.

To start, take one stitch, then tug on your top thread to draw the yarn to the top. To end a line of stitching, cut the top thread and yarn about 4" long. Then pull the yarn to the wrong side with a tapestry needle and tie it together with the top thread. Also tie together the two threads at the start of your stitching.

Use a straight stitch and move your fabric at a consistent pace, making medium length stitches (approximately 8 to 10 per inch). If your stitches are too close together, your bobbin yarn will "bead," and if the stitches are too long, you will see puckers. Make certain your top tension is adjusted properly: if it is too loose, you will be able to see your top thread from the right side. If your bobbin tension is too loose, your bobbin yarn will form sloppy loops on top. When the tension is correct, the bobbin yarn will lie nice and flat on the top of the fabric and the back

side will look like neat top stitching.

The great advantage to bobbin drawing, as opposed to couching with similar yarns, is that you can "draw" very tight loops and circles since you are working free-motion. It would be extremely time consuming to use a foot and pivot around such intricate maneuvers. Also, bobbin drawing is extremely fast. You will become hooked in no time, and think of all kinds of ways to add extra texture to your machine art with the specialty threads it enables you to use. Shown on the opposite page is a sampler made using many different yarns in the bobbin (the thread and yarn ends have been left attached on purpose, for future reference).

OPPOSITE PAGE: A sampler of bobbin drawings with various yarns.

THIS PAGE
Bobbin drawings used in two ways:

RIGHT, TOP: Lorelle LaBarge cleverly follows the evenly spaced dots printed on her reversible fabric to bobbin draw perfectly straight rows of intersecting lines onto her bolero jacket.

RIGHT, BOTTOM: Jeanie Sexton bobbin draws metallic yarns in sprawling loops and twisting curves on her bolero jacket "Lady in the Sun."

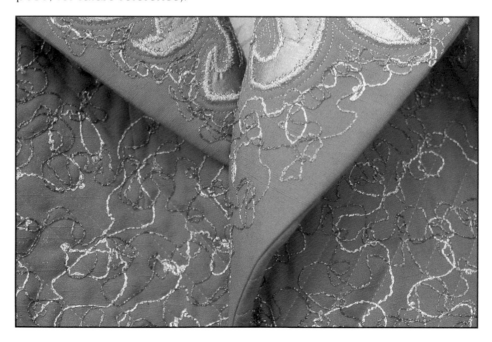

ULTRA-APPLIQUÉ

This is an exciting method incorporating free-motion straight stitching with bobbin drawing. Ultra-suede® is a unique fabric in that it is neither woven nor knit; it is processed by felting which creates a textile that will not fray when cut. Ultra-suede® can be appliquéd with a straight stitch, without the edges raveling. Plus, its rich suede-like surface creates an instant luxurious appeal. Try the following exercise:

Materials: Ultra-suede® scraps in several different colors, firmly woven background fabric, freezer paper, chalk, appliqué scissors, Candlelight™ metallic yarn, thread to match.

MACHINE SET-UP (FOR APPLIQUÉ)
1. Lower feed dogs.
2. Put on the darning foot.
3. Use regular sewing or rayon thread on top to match or blend with Ultra-suede™ scrap.
4. Use regular sewing thread in the bobbin, to blend with top thread. It does not have to match exactly.
5. Use normal top and bobbin tension.
6. Fuse freezer paper to the wrong side of your background fabric.

LEFT: Working from the wrong side, metallic yarn was bobbin drawn along the satin stitching of these appliquéd leaves, to add sparkle.

APPLIQUÉ

This technique is fun, fast and creative, and is best not planned out in advance. With chalk, draw a simple shape on the right side of an Ultra-suede® scrap; it can be a leaf, a simple flower, butterfly, or geometric shape. Draw this simple shape freehand, and do not worry about making a recognizable rendition of anything! Position this scrap of Ultra-suede® where you would like it on your background and (optional) pin it in place.

Now, simply stitch all the way around your chalked outline using tight stitches (approximately 16 to 20 per inch). Remember, you are working free-motion and don't need to pivot! Using your machine appliqué scissors, trim excess Ultra-suede® right next to your stitching. Repeat these steps until you are pleased with your composition. Shapes may overlap – just be very careful that when trimming the top Ultra-suede® shape you do not cut into anything underneath.

Now you are ready to add pizazz with bobbin drawing!

RIGHT: Ultra-suede® leaves cascade down this rayon tapestry vest. They were sketched on scraps of Ultra-suede® and free-motion straight stitched to the background. The Ultra-suede® was trimmed next to the stitching and embellished by bobbin drawing with metallic yarns.

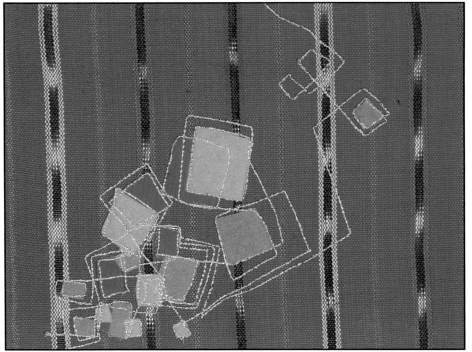

MACHINE SET-UP
(FOR BOBBIN DRAWING)

1. Fill bobbin with Candlelight™ yarn and slightly loosen the bobbin tension.
2. From the wrong side, you will see the straight stitched outline perforating the freezer wrap paper. You can either follow this outline as illustrated in the photo, top left; echo outline it as shown in the photo, bottom left; stitch inside of it (the veins inside of the leaves in the photo, opposite page, top); or make up whole new outlines (photo opposite page, bottom).

I also like to add bobbin drawing after I have appliquéd a design with a satin stitch by following the outline of my appliquéd shapes from the wrong side (again, through the perforations in the freezer paper). The bobbin drawing with metallic yarn adds a very dramatic sparkle.

LEFT, TOP: The bobbin drawing in gold metallic yarn follows the outline of the appliquéd leaves, while contrasting blue yarn radiates from the center of the flowers.

LEFT, BELOW: These little squares of Ultra-suede® almost seem to be bouncing because of the outline bobbin drawing in gold metallic yarn that echoes each box.

RIGHT, TOP: The gold yarn bobbin drawn inside these leaves creates delicate veins.

RIGHT, BOTTOM: The bobbin drawing of gold and turquoise yarn repeats the pattern of the Ultrasuede® buds and extends well into the background.

7. Free-Motion Appliqué

I had been machine appliquéing and embroidering for several years before it occurred to me to combine the two techniques. I loved being able to utilize gorgeous fabrics to create exciting, textural appliqué designs, and I loved the freedom machine embroidery allowed me in thread painting realistic and intricate designs. When I evaluated my designs for machine art, I would choose to appliqué them if it seemed technically feasible to do so; however, if the design seemed particularly detailed, I knew that maneuvering with my appliqué foot would be too difficult, so I would decide to embroider the design.

LEFT: With the feed dogs dropped, designs have been free-motion appliquéd and then enhanced with embroidery.

When I reached a level where the technical aspects of my projects were no longer challenging, I became very frustrated by the restrictions imposed by appliqué. I wanted to be able to interpret in appliqué anything I could draw. Many of my drawings were extremely detailed, and I resented having to simplify them for the sake of application. I wanted to appliqué, not embroider, such intricate designs as Boston ferns, wispy daisies, or tangled vines. Thread painting is, after all, thread painting, and I sorely missed being able to incorporate my passion for fabric into my designs. But when I used basic appliqué (appliqué foot, feed dogs up) it seemed to take forever to maneuver around these complex shapes; also I disliked the hard edge outline that the satin stitches imposed on my

designs. I wanted my flowers and animals to appear as realistic as possible, more like the images I had worked in machine embroidery.

Ultimately, I developed the technique I call free-motion appliqué: by dropping the feed dogs and working free-motion, I am able to travel around my shapes by sliding as opposed to pivoting. Thus, I am able to slide around tiny curves and sharp points and build much more realistic designs. I am no longer hesitant to include curly ferns and jagged leaves in my drawings!

All free-motion appliqué is worked with the darning foot to eliminate the need to use an embroidery hoop. Therefore, it is necessary to back your background fabric with freezer paper. All other steps, (preparing your fabrics with Wonder-Under™ and

tracing placement lines to the freezer paper) will be the same as for basic appliqué. You will want to use machine embroidery thread to appliqué, either rayon or cotton.

PRACTICE SAMPLER

Use the designs in Figs. 7-1 and 7-2 on pages 108-109 to prepare a practice sampler. Bond the appliqué shapes to a 12" square of firmly woven background fabric. Refer to the photo on page 107 for placement. Fuse a 12" square of freezer paper to your background.

Preparation:
1. Prepare appliqué with Wonder-Under™.
2. Drop feed dogs and put on darning foot.
3. Lower top tension slightly and set zigzag to medium-wide width (2.5mm).
4. Fill bobbin with regular thread (the color is not important).
5. Choose machine embroidery thread to match each of the appliqué shapes.
6. Fuse freezer paper to the background.

THIS PAGE: Working free-motion, sharp points and tight curves are conquered easily and beautifully without your having to pivot at all.

OPPOSITE PAGE: Appliqué worked free-motion. Notice that the curves are satin stitched and the points are side stitched; this is achieved by lining the needle to swing into each curve or point.

The direction that you stitch is critical. The sewing machine will form satin stitches when the fabric is pulled forward or backward, and side stitches when the fabric is pulled to the left or the right. For the smoothest result, curves should be satin stitched and points should be side stitched. Therefore, it is important to line your needle swing up with any tiny curve or sharp point before you start stitching. Think of your needle as forming a horizontal line as it swings back and forth. Just as in embroidery, you want the needle to swing into the curve or the point, so line your shape up horizontally. Once you travel around one maneuver, pause and consider whether it is necessary to pivot before approaching the next maneuver. For example, as you are appliquéing the points of your ferns, you will pivot slightly after stitching each point so that the needle swings directly into each one.

With the feed dogs out of commission, you will be in control of stitch length; that is, how close together the stitches will be. If you press your foot pedal faster, shorter stitches will result. Just as in machine embroidery, control will come with practice. I find that a steady speed will form smoother stitches, and I sew satin stitches much faster (pulling my fabric forward or backward) than I sew side

stitches (pulling my fabric sideways).

You will find that as you sew sideways the stitch that is formed will not be wide enough to secure the fabric edge. It will be necessary to move back and forth darning the edges until the stitching is wide enough to secure the raw edge. Again, with practice, side stitching will blend smoothly into the satin stitches.

Free-motion appliqué will look different from a design satin stitched with the appliqué foot. The stitching will widen naturally from narrow side stitching to slanted satin stitches to wider satin stitches as you move the fabric along. No longer will rigid satin stitches outline each shape: instead this beautiful variation of stitch widths will lend a more rounded appearance to the appliquéd design, as shown in the flower pictured above.

After you have completed the flower and fern design, you should feel quite comfortable appliquéing free-motion, and you will continue to gain skill in pulling your fabric at a consistent rate to form perfect satin stitches. You will also start to line your design up in the correct position automatically, without having to evaluate each maneuver. In other words, this technique will start to become very easy and you will be astounded at how simply you can appliqué even the most intricate designs.

Figure 7-1

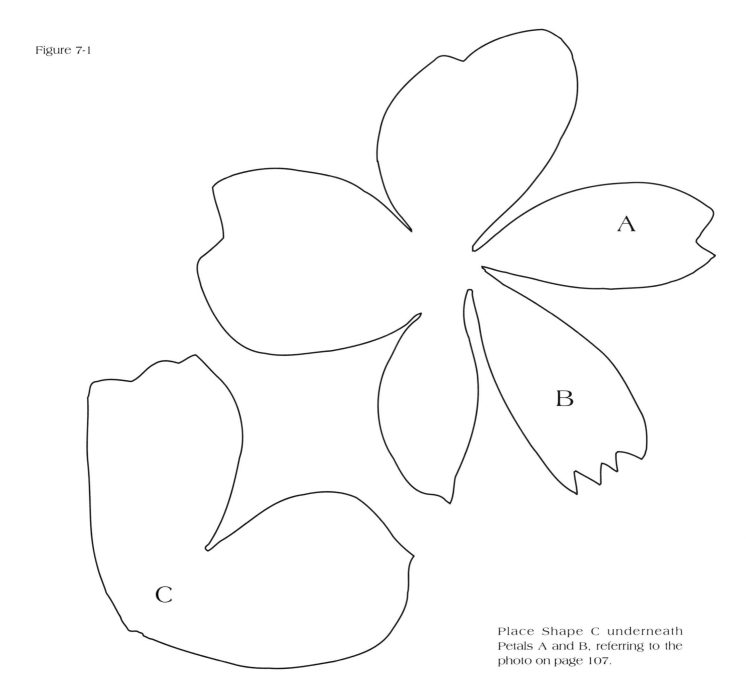

Place Shape C underneath Petals A and B, referring to the photo on page 107.

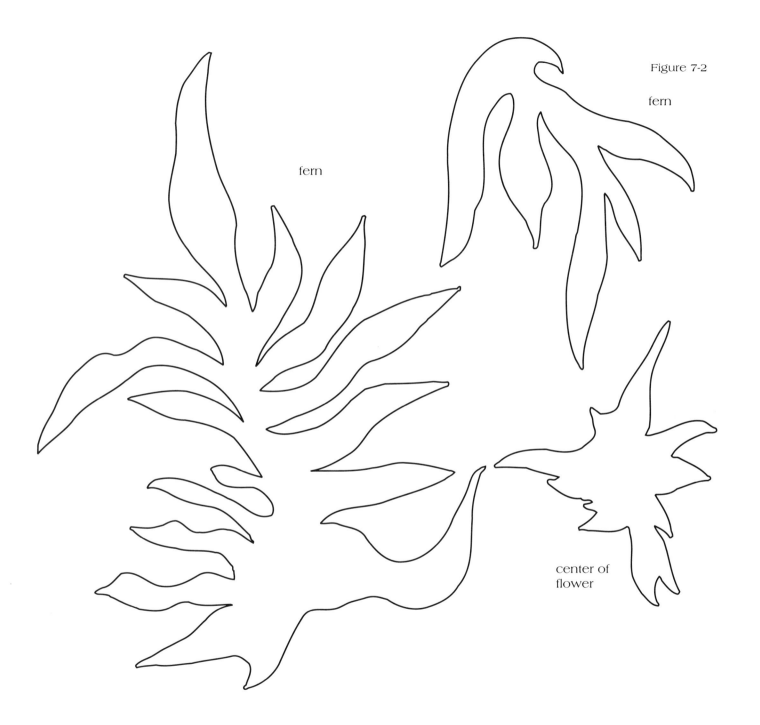

Figure 7-2

fern

fern

center of
flower

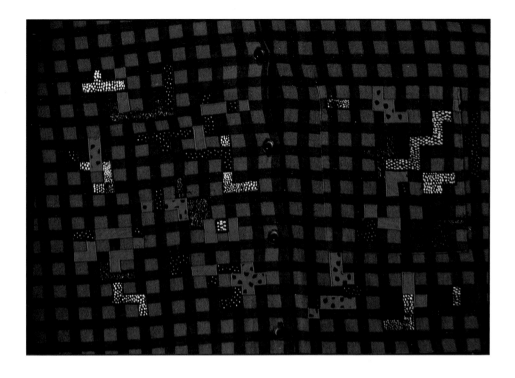

Do I ever use my appliqué foot anymore? Yes! When I am appliquéing very straight, long edges, it is easier to use my machine's feed dogs and the appliqué foot. For example, when I appliquéd the squares and long rectangles on the shirt front in the photograph at left, I used my appliqué foot. I also use it when I want the crisp, heavy outline that results from using traditional appliqué, as in the silk dress appliquéd with Ultra-suede® flowers and couched vines (detail shown below left). The flowers were papercut in a graphic style that looked better outlined with rigid stitches. But whenever your design includes tight curves, sharp points, tiny details and organic shapes, free-motion appliqué will be by far the best method to use. And, it is possible that for some designs it would be more appropriate to use both free-motion and traditional appliqué. You will then be able to say, with smug satisfaction, I can appliqué anything!

TOP: Long straight lines and crisp corners are easier to appliqué using an appliqué foot with the feed dogs engaged.

BOTTOM: These flowers were also appliquéd with the feed dogs engaged. The resulting crisp outline is appropriate for the stylized design.

SHADED APPLIQUÉ

Let's now examine one of the most exciting techniques you can do with your sewing machine: shading appliqué designs with embroidery stitches so that the images appear as realistic as they did with thread painting. And, with shaded appliqué you will have the added pleasure of being able to incorporate textures from your beautiful (and growing!) fabric collection.

Use the design in Fig. 7-3 on page 112 and prepare a 12" square of background fabric, and appliqué fabrics for the flowers and leaves. (Refer to the directions on page 43 of Chapter

Four.) When using the light box, don't forget to trace the detail lines onto the top of your flowers. Also, trace the vines onto your background fabric with a water soluble pen. You will machine embroider them with a side stitch. The photo at bottom left shows some of the appliqué shapes being positioned.

You are now ready to free-motion appliqué. It is critical that you select a thread color to perfectly match each appliqué shape. As you are appliquéing, include the detail lines that you have traced. Your design will look like the one shown in the photo at bottom right when you are finished.

This is a beautiful flower, but

it appears rather flat, doesn't it? So let's add some three-dimensional realism!

To understand how to shade in embroidery, it is necessary to

BOTTOM LEFT: Placement Lines: The mirror image of the design has been traced to the freezer paper and the appliqué shapes have been traced, fused to the fabric, and cut out. All detail lines (see flower example) have been traced to the right side. The freezer paper is bonded to the background and the appliqué shapes positioned over the placement lines and fused in place.

BOTTOM RIGHT: Free-Motion Appliqué: Match thread as closely as possible and free-motion appliqué all shapes; include detail lines (see petals of flower). Peel away freezer paper when completed.

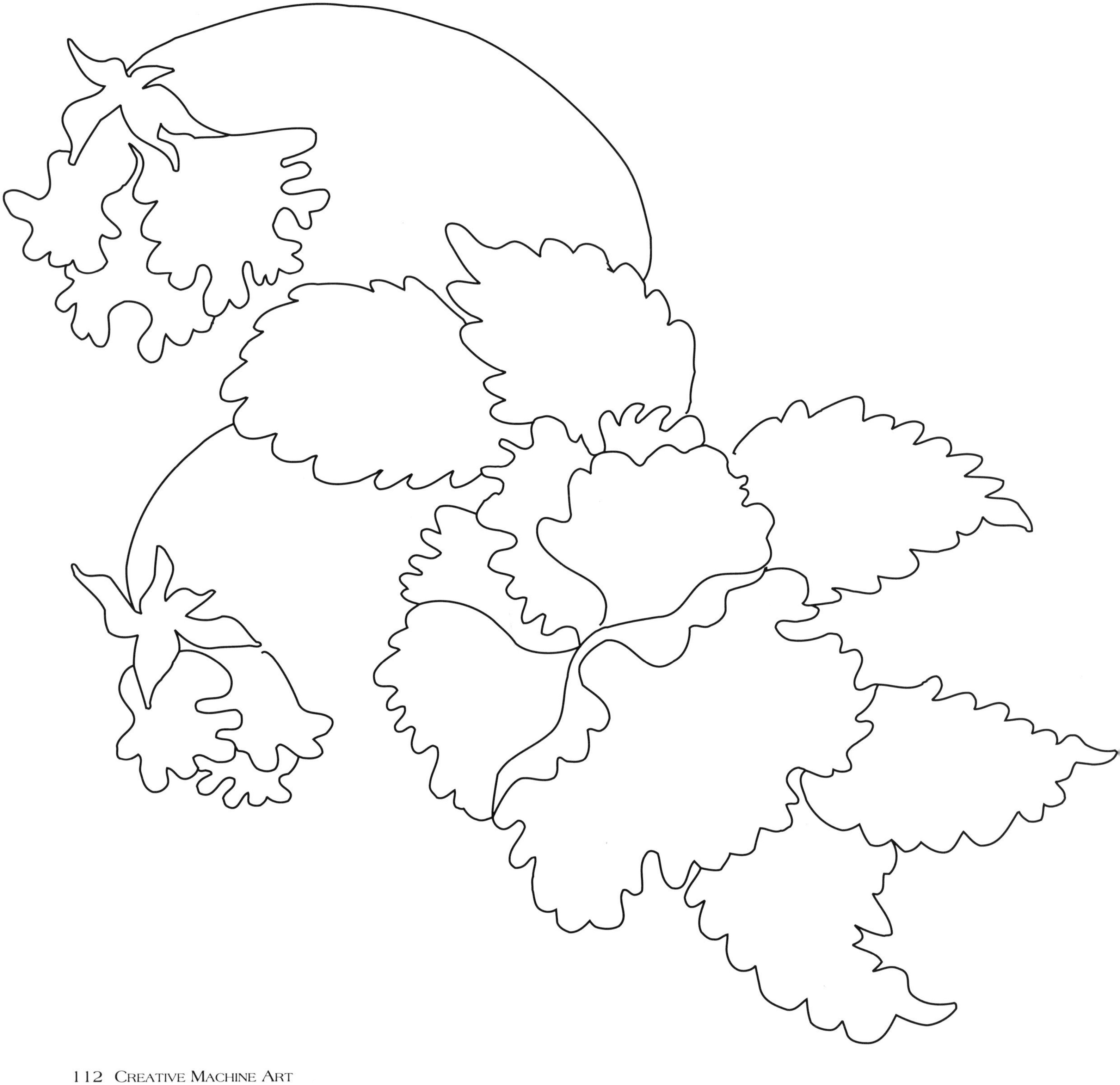

have a good understanding of how value creates depth. For example, light values will advance while dark values will recede. Therefore, the closest part of a design will be a lighter value than the part farthest away. Also, when areas of a design overlap, the part underneath will be in shadow while the area on top will be lighter.

Since it is necessary to think in terms of black and white (and gray) to successfully plan a shaded design, I carefully plan my values before I plan my colors. I reduce my design to a black and white sketch so I am not influenced by any colors. Then I use colored pencils in four graduated shades of gray to color in my design. (Berol™ Prisma® colors #935, #965, #966 and #967 work well, as do other brands of colored pencils in graduated shades.) These areas of shading from my sketch will be used as a guideline for stitching with corresponding values of thread onto my fabric.

RIGHT: Shaded appliqué creates an irresistible llama with charisma! Embroidered straight stitches are used to add shading and detailing, helping shape his winsome personality.

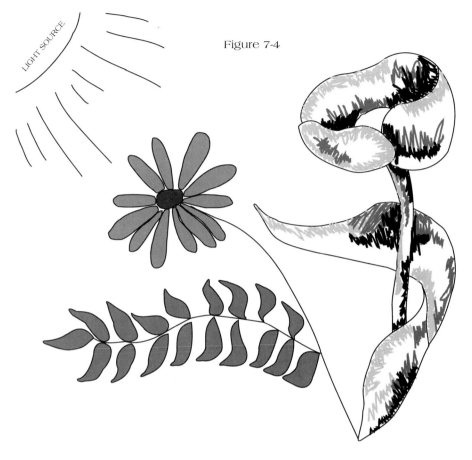

Figure 7-4

SHADING TIPS

- I find it very helpful to imagine a light source. If I can think of my design as being lighted from a particular area, then it is easier for me to determine where to highlight and add my lightest values. I simply draw a sun in a position on my sketch. It can be anywhere – off to one side, above, or below! Then I think of the rays coming from the sun to my design. Highlights will be the areas of the design closest to this light source. (See Fig. 7-4, at left, and corresponding photo below.)

- With my lightest gray pencil, I color in areas that I imagine are highlighted. These will be those areas highlighted by my "sun" and areas of the design that are overlapping others (for example, the stem that is in front of a leaf). I press down as I color because I want to be sure to be able to distinguish this gray pencil from subsequent ones.

- Next, I switch to my darkest pencil (black) and color in areas that I feel will be shaded. These will be the parts that are farthest from the light source and those parts that are in the shadows caused by overlap-

LEFT: Drawing used for planning shading and the shaded appliqué created using the drawing as a guide. The graduated shades of thread simply replace the graduated shades of pencil!

ping (for example, the leaf behind a stem).

- After defining the highlights and the shades, middle values of gray will be used to blend. I use the darker grays to blend into the black, and the lighter grays to blend into the high-lighted areas.

This drawing is my road map to guide me while stitching. The next step is the most critical, and that is selecting the thread. Again, I think in terms of value as opposed to color. First of all, I consider the thread that I used to appliqué my design as my middle value. In other words, one of the grays used for blend-ing. Then I carefully pick out val-ues of that same color that are darker and lighter. The more gra-dations of value (the more spools of thread) I select, the more shaded my design will appear; I like to use at least four shades. I then assign them "light," "medium," or "dark" refer-ring to my shaded drawing, and line them up accordingly. All that's left is to color in with my sewing machine, matching the values of my thread with the val-ues of my drawing!

RIGHT: Paulette Thompson has enhanced these Ultra-suede® roses with shaded appliqué to embellish this gorgeous wool crepe suit.

ADDING THE MACHINE EMBROIDERY

This step, usually the one my students anticipate with trepidation, is actually the easiest part. The shades have already been assigned their location according to the shaded drawing. All that is left is to apply them with thread where they have been shaded with the colored pencils. The technique of applying the thread by machine is easy! By using free-motion techniques and a straight stitch, there is no need to pivot or to be concerned with which direction to stitch. It is exactly like drawing, or coloring with a pencil, with one exception. Since it is not possible to move the sewing machine as you would your pen-

cil, it will be as though you are moving your paper beneath the crayon – or your fabric beneath the needle. Some people color with tiny crosshatches, some people color with loops or circles, and some people color with broad strokes. Everyone's style of coloring will vary, and the drawings will also. The same holds true for the shaded embroidery, and I encourage my students to develop their own styles of stitching instead of trying to reproduce mine, for there is no right or wrong way to embroider. What is important is to achieve the feeling of depth, and that depends on where the lights and the darks are placed, not how they are stitched.

Set your sewing machine up as you did for the free-motion appliqué: drop the feed dogs, put on the darning foot, and lower the top tension slightly. You will be using the straight stitch setting, so there are no further adjustments for width or

OPPOSITE PAGE
LEFT & BOTTOM RIGHT: Jeanie Sexton's expert application of shaded appliqué transforms fabric and thread into these exquisite Oriental girls gracing "Fan-tasy Dance." The close-up illustrates how effective thread is when embroidered on top of fabric to add detail and texture.

RIGHT: Peach tinted poppies adorn Jeanie Sexton's elaborately appliquéd and shaded wool crepe suit, "Playful Poppies."

length. As with the free-motion appliqué, the faster you press the foot pedal, the shorter your stitches will be, and the faster you move the fabric, the longer the stitches will be!

I like to add the embroidery in the same sequence that I colored in the shades with my gray pencils: starting with my lightest value, then adding my darkest value, then blending the values together with my medium shades.

Look at the photo, page 118, top left. The lights of each color (red for the flower and green for the leaves) have been added. In the top right photo on page 118, the darks of each color have been added. At this point the design will not look pretty. The

lights and darks appear too exaggerated, but this much contrast is necessary to create the illusion of depth. It is important to remember that the grays are going to soften and blend these strong values.

In the bottom left photo on page 118, the grays have been added. Use the lightest grays to blend next to the highlighted areas, and the darkest grays to soften the strong darks. Part of the shaded stitches will mix in with the dark and light stitches, and part of them will extend over the appliqué satin stitches.

In the bottom right photo on page 118, the thread that was used to appliqué each shape has been added to further soften and blend the shaded areas.

Also, I have added a straight stitch to outline each appliqué shape. I chose a color that was a dark shade of the appliqué shape and free-motion straight stitched right next to the satin stitching; this seemed to define each shape and add dimension. This step is also very effective if batting is placed underneath (remove the freezer paper first!) so that the design is quilted as you stitch. In the example shown, I have added some bobbin drawing with metallic yarn for the trailing vines.

You can always add more stitches after evaluating your piece. You can even "cover up" any stitches that you are not pleased with by embroidering over them with a different shade. It is easy, though, to get carried away and add too much embroidery; you may inadvertently cover up your beautiful appliqué fabric! Remember that the embroidery is added only to shade your design and too little is always better than too much, because you can go back and add more if needed. As you are embroidering, it is important to view your design at a distance every once in a while: either hold it at an arm's distance away or pin it to a wall and step back. The values will stand out more when the appliqué is viewed from a few feet away and this will make it easier to distinguish what values are needed where.

OPPOSITE PAGE
TOP LEFT: HIGHLIGHTS: Stretch the fabric in a machine embroidery hoop and remove your darning foot. Refer to your shaded drawing and add all lights of each color.

TOP RIGHT: DARKS: Again referring to drawing, add all darks of each color. At this point the design will not look pretty. The lights and darks will appear too exaggerated, but this much contrast is necessary to create the illusion of depth. It is important to remember that the grays are going to soften and blend these strong values.

BOTTOM LEFT: GRAYS: Use the lightest grays to blend next to the highlighted areas, and the darkest grays to soften the strong darks. Part of the shaded stitches will mix in with the dark and light stitches, and part of them will extend over the appliqué satin stitches.

BOTTOM RIGHT: FINISHING: The thread that was used to appliqué each shape is added to further soften and blend the shaded areas. I like to add a straight stitch to outline each shape. For this design, I chose a color that is a dark shade of the appliqué shape and free-motion straight stitched right next to the satin stitching. On this example, I have also added some bobbin drawing with metallic yarn for the trailing vines.

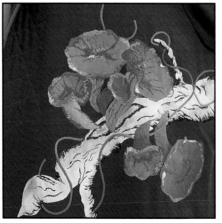

After all stitching is completed and you are satisfied with how your design looks, remove all paper from the back. Large appliquéd areas will tear away easily, and the tiny bits of paper caught between the embroidery stitches can be removed with tweezers.

Optional: You may prefer to tear the paper away after you have completed the free-motion appliqué, but before you add the shaded embroidery stitches. Stretch your fabric in a hoop to add the free-motion straight stitches. Then you will not have to deal with removing any paper underneath the embroidery.

If there is any puckering, your design will need to be blocked. Place your design, right side down, on a padded ironing board and spray it with a fine mist of water from your steam iron or an atomizer. Then smooth away the puckers with your hands and carefully press with an iron.

This combination of applying machine embroidery with appliquéd fabrics will add a whole new dimension to your designs. You, too will become convinced that anything that can be drawn or painted can be reproduced with fabric and thread!

OPPOSITE PAGE: Judy Simmons masterfully executes shaded appliqué to create a forest floor teaming with color and texture in "Mushroom Medley."

TOP: Karen Ballash has so skillfully shaded her appliquéd underwater scene that it is easy to visualize sea water swishing through the seaweed and coral.

BOTTOM: Hummingbirds are captured in flight in Paulette Thompson's masterful rendition of her shaded appliqué class project. Paulette has cleverly made a matching purse to go with this exquisite wool crepe dress.

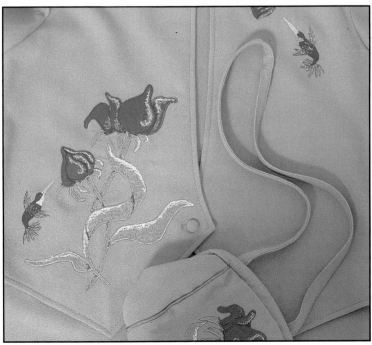

FACING PAGES: Linda Stallion evokes deep feelings of patriotism with her beautifully appliquéd cape depicting symbols of the United States. A close-up of the majestic eagle reveals Linda's expert combination of free-motion appliqué and embroidery.

8. *Duplicating Children's Art*

Kids are so creative! I've always been a strong advocate of encouraging children to draw because they are able to express themselves very naturally, and also seem to have an innate sense of good design. Often times, children lose some of their spontaneity as they grow older. As adults, we start to impose restrictions on ourselves. As we struggle to master technical skills and to create within the bounds of formalized design, we easily lose sight of what it was that encouraged us to create in the first place!

OPPOSITE PAGE: "My World," detail, was designed by Danielle Roberts when she was nine years old. Each block depicts people, pets and places that are important to her. Machine art skills can be used to preserve children's imaginative drawings in fabric and thread.

But children seem to draw as naturally as they breathe, and the effect is that they can capture a mood, an emotion, a memory or dream that each of us, as viewers, can identify with in our own way. Isn't that what being a good artist is all about? I have always been secretly jealous of the ease with which my two children can sit down and create an absolute masterpiece in a matter of minutes. I can examine their drawings every day and each time find something new to smile and wonder at. They see the world differently, and their drawings help us remember life as a child.

Machine art is the perfect medium for preserving these very fragile recordings. As carefully as I try to save my children's special drawings, they always seem to become torn, bent, or lost. By combining appliqué with embroidery, I can reproduce my children's artwork in exact duplicate, using fabric and thread to add color, texture and durability. I like to have my children choose the fabrics to be used for their appliqué; this makes the final project very much their own creation. We machine artists are merely the technicians, and by using our skills we can create a very lasting heirloom of our children's creative talent.

ANALYZING THE ART AND PREPARING THE DESIGN

It is important to make decisions as to what techniques will be used for reproducing the drawing. Since free-motion appliqué will be used for most of

OPPOSITE PAGE: "My World," 80" x 96". A few of the highlights: *ROW 1.* **BLOCK 1**: Danielle snuggles beneath her quilt as she spies on the Easter Bunny leaving behind a basket full of goodies. *ROW 3.* **BLOCK 3**: Danielle's aunt runs a bridal shop, which apparently is a romantic and delightful place. Note the bridal gown hanging on the rack, appliquéd in bright yellow daisies. *ROW 4.* **BLOCK 1**: Chapped cheeks are ringed in red and pink to indicate the freezing temperature of this winter day. Notice that each embroidered snow flake is straight stitch outlined with dark blue thread. **BLOCK 3**: It must surely seem to Danielle that her body is forming a perfect circle as she demonstrates a backward flip to a proud Granny. *ROW 5.* **BLOCK 2**: Patches the cat is immortalized along with her offspring, Cricket. **BLOCK 3**: Even the squirrel is smiling to indicate the joyful abandon Danielle and her cousin experience when they jump into a pile of crisp (and very colorful!) autumn leaves. **BLOCK 4**: Simple pleasures are best: riding in Pop's pick-up with her cousins rates as a favorite memory for Danielle. *ROW 6.* **BLOCK 2**: Summer is so carefree even fishing lines never tangle; rather, they conveniently suspend from the sky. **BLOCK 3**: Not even attempting to realistically render the awkward body position, Danielle simply draws the arms on backwards to hide the treat until Beluah the dog obeys the command to sit!

RIGHT: "My World," detail of the fourth block in row 3, Hines Hole. It is evident that the best part of jumping into a favorite swimming hole is the big splash; notice the care with which Danielle colors the sparkling spatters of water.

the appliqué, even tiny images can be cut from fabric and appliquéd. However, very fine detailing will need to be done in machine embroidery. The first step will be to decide what will be appliquéd, and what will be embroidered.

Cut a piece of freezer paper the same size as the child's drawing. Place the drawing *right side down* on a light box. Place the freezer paper, *fusible side down*, on top of the drawing.

Trace the drawing, using an extra-fine point permanent marker.

Use this mirror image drawing to trace any shapes to be appliquéd with Wonder-Under™. Place the Wonder-Under™, *fusible side down*, over the images and trace each shape with your permanent marker. Be sure to trace all of the detail lines within each shape, and add seam allowances to pieces that overlap (see page 43, Chapter 4).

RIGHT: "Life Under the Sea" drawing by Zachary Roberts, age 10.

BELOW: A mirror image of the drawing has been traced onto a piece of freezer paper. The freezer paper is then fused to the wrong side of the background fabric.

OPPOSITE PAGE
TOP, LEFT & RIGHT: When placed on a light box, the placement lines clearly show through. All appliqué pieces are positioned and fused in place. Any detail lines that are to be embroidered are traced with a fine pointed, permanent black pen.

BOTTOM, LEFT & RIGHT: The embroidery has been added to complete the design.

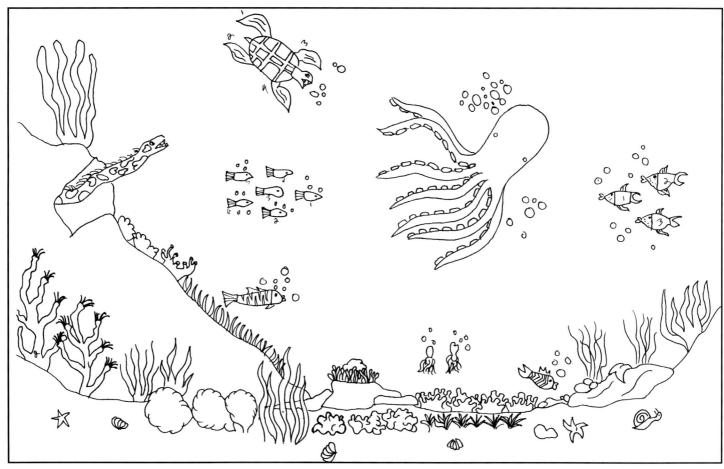

Sometimes you will appliqué part of a design and embroider the tiny details. The decisions as to which parts will be appliquéd, and which will be embroidered must be made as you are preparing your appliqué shapes. If there are any parts inside an appliqué shape that will be embroidered, they must be traced on top of the appliqué shape with a permanent marker.

Look at the sea turtle in "Life Under the Sea" (photo bottom left). His legs and shell are appliquéd, but it was simpler to embroider the design inside his shell. After the turtle shell was cut out of fabric, it was placed over a light box with the Wonder-Under™ still attached to the underside. The shell design that has been traced onto the Wonder-Under™ showed through the top, and it was then traced with the marker. Be sure to trace these detail lines *before* removing the Wonder-Under™.

Now look at the tiger striped fish (top right photo); the main part of his body is appliquéd, but the fins, head and tail were too tiny to cut out of fabric so they will be embroidered. The embroidered part of this shape extends beyond the appliquéd part, so it must be traced with the marker. Remember, the entire design is traced on the freezer paper underneath, so all placement lines will show very clearly when the background is

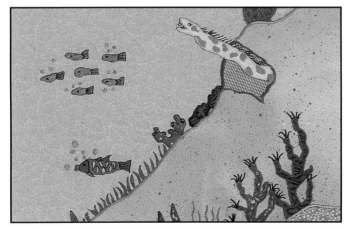

placed over the light box. I fused the tiger fish's body in place (after tracing the tiger stripes), and then, with the marker, I traced the head, fins, and tail to be embroidered.

There are no rules for determining which shapes to appliqué, and which to embroider. I like to incorporate fabric whenever I can, so I will appliqué even very small shapes. The appliqué will be worked free-motion; sometimes using a very narrow stitch width if the shape is tiny. With practice, even ½" shapes can be appliquéd! I think most designs look better if they are worked in a combination of appliqué and embroidery; I suppose I just like textural contrast of fabric against thread. However, these are personal choices, and ones that each machine artist makes on the basis of preference and skill level.

When appliqué designs have multiple layers, the tracing on the freezer paper will not show through all of the layers of fabric. Sometimes it is possible to fuse appliqué shapes to sections before fusing those sections to the background. For example, in "Life Under the Sea," the red and purple plants were fused to the bottom of the red boulder before that boulder was bonded to the background. That way, the plants could be positioned using the traced lines of the Wonder-Under™ from the boulder. If this

is not possible, you will need to trace any placement line of layered shapes with a water soluble marker. Do not use your permanent marker, because the shape may not cover all of the lines exactly, and the water soluble lines can be erased.

APPLIQUÉING AND EMBROIDERING THE ART

Trace all detail lines and bond all appliqué shapes in position until your picture includes everything your child has drawn; the picture should be an exact reproduction. Look at the pictures, "Life Under the Sea" (page 129) and "Sea Monster" (page 131) to see how they looked

before any stitching was added.

At this point, most of the work has been completed! The stitching goes much faster than the preparation because most of it is worked free-motion. The only appliqué that I use my foot for is long lines. For example, the boulders in both pictures were appliquéd using my foot, as it is easier to let my feed dogs push the fabric along for these lines, to obtain a consistent satin stitch. Everything else was stitched with the feed dogs dropped and a darning foot attached (see "Free-motion Appliqué"). I switched easily from embroidery to free-motion appliqué whenever the design dictated a change, since I didn't need to use a hoop for the

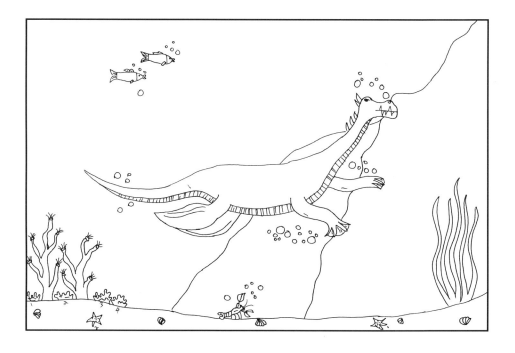

RIGHT & OPPOSITE PAGE: The "Sea Monster" drawing by Zachary Roberts, age 10, (shown right) was the basis for the block shown below. A mirror image (opposite page) of the drawing was traced on to freezer paper.

CENTER, LEFT & RIGHT: Fusing and tracing complete; ready for stitching.

BOTTOM, LEFT & RIGHT: Appliquéd and embroidered. Even Zachery's finest details can be duplicated just as he drew them. Note the multi-colored bubbles and the webbed feet, striped underbelly and pink spikes of his snaggle-tooth sea monster.

embroidery.

Don't forget to use that handy straight stitch in your embroidery! It is especially helpful in adding tiny details, such as the wispy plants along the ocean floor. Also, it is very useful in outlining appliquéd shapes to help define them. Look at the Moray eel poking his head from his hiding hole in the "Life Under the Sea" block on page 129. If I had appliquéd him using black thread, the inside of his body would have disappeared because he is so very small. If I had simply appliquéd him in matching thread, he would not have shown up. So I appliquéd him in the white thread, and then stitched along the outside of the satin stitching using black thread and a straight stitch. This outline straight stitch is very useful, and can be worked in other colors as well.

Duplicating my children's artwork is probably the most challenging machine art I have done. I remember looking in alarm at my daughter's first block drawings for her quilt, "My World." I contemplated all the tiny details she had included and wondered how I was going to appliqué them. I have a horrible confession to make: I asked her to draw one of the blocks again, with fewer details and larger figures. When she completed the second drawing, I instantly felt ashamed. I had imposed my standards on my child, and she was no longer drawing from her own imagination. I was looking at a "dead" drawing, for the magic that had made all of her other drawings so special was missing! I learned my lesson, and improved my machine art techniques to be able to reproduce whatever my children created. I have been able to incorporate in my own designs the skill and technique I have gained. After you have successfully re-created a child's drawing in fabric and thread, you will be able to reproduce any design!

BELOW, LEFT & RIGHT: Riding a pony for the first time made a memorable and happy impression on Danielle. Notice how effective a straight stitch in black thread is to outline the tiny hands, bridle reins, and eye glasses.

OPPOSITE PAGE
TOP: Danielle captured the whole meaning of Christmas with her simple sketch, which was re-created in fabric and thread.

BOTTOM LEFT & RIGHT: This two-headed dragon by Zachary, breathes metallic fire and sports jaunty red spots on his marbleized hide. Even the tiniest details can be duplicated; note the dragon teeth and the bumpy spikes down the dragon's back.

For Further Reading

The Complete Book of Machine Embroidery
Robbie and Tony Fanning

Decorating with Machine Embroidery
The DMC Corporation

Creative Machine Embroidery (Volumes 1, 2, and 3)
Lucille Merrell Graham

Decorative Machine Stitching
Singer Sewing Reference Library

The Art of Embroidery
Julia Barton

Machine Embroidery Lace & See-Through Techniques
Moyra McNeill

The New Machine Embroidery
Joy Clucas

Credits

The garment by Lorelle LaBarge on page 99 was part of a project funded by a grant from the Illinois Arts Council and Southern Illinois Art Expansion Art/Access Program.

Traditional quilt blocks used for examples in Chapter 5 are taken from:

A Quilter's Companion
Dolores A. Hinson
Indian Wedding Ring, pp. 91

Crib Quilts and Other Small Wonders:
Thos. K. Woodard and
Blanche Greenstein
Carolina Lily, pp. 118

A Second Quilter's Companion
Dolores A. Hinson
Susan's Wreath, pp. 88

Suppliers

WEB OF THREAD

3240 Lone Oak Road, Suite 124

Paducah, KY 42003

Phone: (502) 554-8185

Fax: (502) 554-8257

Speciality threads and yarns; Sulky™ rayon machine embroidery thread metallic, nylon, and silk threads, yarns, ribbons, cords and trims, Perfect Pleater™, embroidery hoops.

Catalog: $2, refundable with first order.

ENGLISH'S SEWING MACHINE COMPANY

7001 Benton Road

Paducah, KY 42003

(800) 525-7845

Mez Alcazar™ rayon thread, DMC™ cotton embroidery thread, stabilizers, Wonder-Under™, darning feet, water-soluble markers.

NANCY'S NOTIONS

33 Beichl Avenue

P.O. Box 683

Beaver Dam, WI 53916

Nylon and metallic thread, hoops, Wonder-Under™, marking pens.

SPEED STITCH

3113 Broadpoint Dr.

Harbor Heights, FL 33983

1-800-874-4115

Sulky™ rayon machine embroidery thread, metallic and nylon thread, hoops, stabilizers, Wonder-Under™, Sharpie™ pens.

CLOTILDE

1909 S.W. First Avenue

Fort Lauderdale, FL 33315-2100

1-800-545-4002

Perfect Pleater™.

Photo

Page 56 bottom	TRAILING STARS	Photo: Elizabeth Courtney
Page 57 bottom left		Photo: DSI Graphics, Donahue Studios Inc., Evansville, IN
Page 57 bottom right		Photo: DSI Graphics, Donahue Studios Inc., Evansville, IN
Page 58 left		Photo: DSI Graphics, Donahue Studios Inc., Evansville, IN
Page 58 center	TIMELESS SHADOWS	Photo: Elizabeth Courtney
Page 58 right	TIMELESS SHADOWS	Photo: Elizabeth Courtney
Page 59		Photo: DSI Graphics, Donahue Studios Inc., Evansville, IN
Page 60 top	DIAMOND ICICLE	Photo: Elizabeth Courtney
Page 60 bottom left		Photo: DSI Graphics, Donahue Studios Inc., Evansville, IN
Page 60 bottom right		Photo: Elizabeth Courtney
Page 61	SPIRITS OF THE NIGHT WIND	Photo: DSI Graphics, Donahue Studios Inc., Evansville, IN
Page 62 series		Photo: DSI Graphics, Donahue Studios Inc., Evansville, IN
Page 63	LET'S GO TO MARKET	Photo: Elizabeth Courtney
Page 64	CAROLINA LILY Block	Photo: DSI Graphics, Donahue Studios Inc., Evansville, IN
Page 65 bottom left	INDIAN WEDDING RING Block	Photo: DSI Graphics, Donahue Studios Inc., Evansville, IN
Page 65 bottom right	INDIAN WEDDING RING Block	Photo: DSI Graphics, Donahue Studios Inc., Evansville, IN
Page 66-67 series		Photo: Sharee Roberts
Page 69	SUSAN'S WREATH Block	Photo: DSI Graphics, Donahue Studios Inc., Evansville, IN
l		
Color Plate i	GUARDIAN OF THE ENCHANTED DIAMONDS	Photo: Brad Stanton, Fairfield Processing
Color Plate ii	CHERRY BLOSSOM	Photo: Glenn Hall, model Joanna Pace
Color Plate iii	SUNSET OVER DIAMONDHEAD	Photo: Brad Stanton, Fairfield Processing
Color Pate IV	TIMELESS SHADOWS	Photo: Elizabeth Courtney
Color Pate v	TROPICAL RENDEZVOUS	Photo: Glenn Hall, model Joanna Pace
Color Plate vi	TROPICAL RENDEZVOUS	Photo: Glenn Hall, model Joanna Pace
Color Plate vii	SPIRITS OF THE NIGHT WIND	Photo: Glenn Hall, model Joanna Pace
Color Plate viii	MONKEY ON MY BACK	Photo: Glenn Hall, model Joanna Pace
Color Plate ix	AUTUMN GLORY	Photo: Glenn Hall, model Joanna Pace
Color Plate x	DIAMOND TRELLIS	Photo: Elizabeth Courtney
Color Plate xi	SNAKES IN THE GARDEN	Photo: Glenn Hall, model Joanna Pace
Color Plate xii	THE BEAST WITHIN ME	Photo: Glenn Hall, model Joanna Pace
Color Plate xiii	PATH BY THE FALLS	Photo: Glenn Hall, model Joanna Pace
Color Plate xiv	MYNA RIOT	Photo: Glenn Hall, model Joanna Pace
Color Plate xv	NEW YORK RED	Photo: Glenn Hall, model Joanna Pace
Color Plate xvi	LET'S GO TO MARKET	Photo: Elizabeth Courtney
Color Plate xvii	UNTAMED SPLENDOR	Photo: Glenn Hall, model Joanna Pace
Color Plate xviii	UNTAMED SPLENDOR	Photo: Glenn Hall, model Joanna Pace
Page 72		
Page 74	ICY WINTER STORM/Marilyn Boysen	Photo: Bill H. Boysen
Page 79 bottom left		Photo: DSI Graphics, Donahue Studios Inc., Evansville, IN
Page 79 bottom right		Photo: DSI Graphics, Donahue Studios Inc., Evansville, IN
Page 80 top left		Photo: DSI Graphics, Donahue Studios Inc., Evansville, IN
Page 80 top right		Photo: DSI Graphics, Donahue Studios Inc., Evansville, IN
Page 81		Photo: DSI Graphics, Donahue Studios Inc., Evansville, IN
Page 82		Photo: Elizabeth Courtney
Page 83 top left		Photo: Elizabeth Courtney
Page 83 top right	Beverly Burton	Photo: Elizabeth Courtney
Page 86 top	LET'S GO TO MARKET	Photo: Elizabeth Courtney
Page 86 bottom	LET'S GO TO MARKET	Photo: Elizabeth Courtney
Page 87 top	Beverly Burton	Photo: Elizabeth Courtney
Page 87 bottom	TRAILING STARS	Photo: Elizabeth Courtney
Page 88		Photo: Elizabeth Courtney
Page 90		Photo: DSI Graphics, Donahue Studios Inc., Evansville, IN
Page 91 bottom left	MONKEY ON MY BACK	Photo: Glenn Hall

Page 91 bottom right	NEW YORK RED	Photo: Glenn Hall
Page 93 bottom left		Photo: DSI Graphics, Donahue Studios Inc., Evansville, IN
Page 93 bottom right		Photo: DSI Graphics, Donahue Studios Inc., Evansville, IN
Page 94 top		Photo: DSI Graphics, Donahue Studios Inc., Evansville, IN
Page 94 bottom		Photo: DSI Graphics, Donahue Studios Inc., Evansville, IN
Page 95 top		Photo: DSI Graphics, Donahue Studios Inc., Evansville, IN
Page 95 bottom		Photo: DSI Graphics, Donahue Studios Inc., Evansville, IN
Page 96		Photo: DSI Graphics, Donahue Studios Inc., Evansville, IN
Page 97		Photo: Elizabeth Courtney
Page 98		Photo: DSI Graphics, Donahue Studios Inc., Evansville, IN
Page 99 top	Lorelle LaBarge	Photo: Elizabeth Courtney
Page 99 bottom	LADY IN THE SUN/Jeanie Sexton	Photo: Sharee Roberts
Page 100	MOONLIT PATH	Photo: Elizabeth Courtney
Page 101	MOONLIT PATH	Photo: Elizabeth Courtney
Page 102 top		Photo: DSI Graphics, Donahue Studios Inc., Evansville, IN
Page 102 bottom		Photo: DSI Graphics, Donahue Studios Inc., Evansville, IN
Page 103 top		Photo: DSI Graphics, Donahue Studios Inc., Evansville, IN
Page 103 bottom		Photo: DSI Graphics, Donahue Studios Inc., Evansville, IN
Page 104	UNTAMED SPLENDOR	Photo: Glenn Hall
Page 106 top	MYNA RIOT	Photo: Glenn Hall
Page 106 bottom	MONKEY ON MY BACK	Photo: Glenn Hall
Page 107		Photo: DSI Graphics, Donahue Studios Inc., Evansville, IN
Page 108		
Page 108 bottom		
Page 109 series		Photo: DSI Graphics, Donahue Studios Inc., Evansville, IN
Page 110 top		Photo; Elizabeth Courtney
Page 110 bottom		Photo; Elizabeth Courtney
Page 111 bottom left		Photo: DSI Graphics, Donahue Studios Inc., Evansville, IN
Page 111 bottom right		Photo: DSI Graphics, Donahue Studios Inc., Evansville, IN
Page 113	LET'S GO TO MARKET	Photo: Elizabeth Courtney
Page 114		Photo: DSI Graphics, Donahue Studios Inc., Evansville, IN
Page 115	Paulette Thompson	Photo: Rosemary Ponte
Page 116 left	FAN-TASY DANCE/Jeanie Sexton	Photo: Curtis and Mays Studio, Paducah, KY
Page 116 right	FAN-TASY DANCE/Jeanie Sexton	Photo: Curtis and Mays Studio, Paducah, KY
Page 117	PLAYFUL POPPIES/Jeanie Sexton	Photo: Gene Boaz Photography, Paducah, KY
Page 118 series		Photo: DSI Graphics, Donahue Studios Inc., Evansville, IN
Page 120 top	MUSHROOM MEDLEY/Jeanie Sexton	Photo: Judy Simmons
Page 120 bottom	MUSHROOM MEDLEY/Jeanie Sexton	Photo: Judy Simmons
Page 121 top	FISHING FOR COMPLIMENTS/Karen Ballash	Photo: Cheryl Murphy
Page 121 bottom	Paulette Thompson	Photo: Elizabeth Courtney
Page 122	FREEDOM FOREVER/Linda Stallion	Photo: Gene Boaz Photography, Paducah, KY
Page 123	FREEDOM FOREVER/Linda Stallion	Photo: Gene Boaz Photography, Paducah, KY
Page 124	MY WORLD	Photo: DSI Graphics, Donahue Studios Inc., Evansville, IN
Page 126	MY WORLD	Photo: DSI Graphics, Donahue Studios Inc., Evansville, IN
Page 128 top	MY WORLD	Photo: DSI Graphics, Donahue Studios Inc., Evansville, IN
Page 128 bottom		Photo: DSI Graphics, Donahue Studios Inc., Evansville, IN
Page 129 series		Photo: DSI Graphics, Donahue Studios Inc., Evansville, IN
Page 130		Photo: DSI Graphics, Donahue Studios Inc., Evansville, IN
Page 131 top	Art: Zachary Roberts	Photo: DSI Graphics, Donahue Studios Inc., Evansville, IN
Page 131 bottom		Photo: DSI Graphics, Donahue Studios Inc., Evansville, IN
Page 132		Photo: Sharee Roberts
Page 133 series	Art: Zachary Roberts	Photo: Sharee Roberts
Page 139		Photo: Elizabeth Courtney

The Author

Sharee Dawn Roberts, of Paducah, Kentucky, holds a Fine Arts Degree in Textile Design from San Diego State University, and has received both national and international recognition for her skill in machine art. She was the recipient of awards in the American Quilter's Society Fashion Show for three consecutive years, and was the grand prize winner in both 1987 and 1988. She placed first in the Fabric Fantasies Annual Juried Fabric Festival at Bazaar Del Mundo, San Diego, California, in 1990 and 1991. She has been a fashion designer for the Fairfield/Processing Fashion Show for 1988, 1989, and 1990. Her clothing has been shown in galleries and exhibitions throughout the United States, Japan and Europe. Sharee is also nationally recognized as an outstanding teacher and lecturer in her fields of machine art and fashions.

❧ American Quilter's Society ❧

dedicated to publishing books for today's quilters

The following AQS publications are currently available:

- **Adapting Architectural Details for Quilts,** Carol Wagner, #2282: AQS, 1991, 88 pages, softbound, $12.95
- **American Beauties: Rose & Tulip Quilts,** Gwen Marston & Joe Cunningham, #1907: AQS, 1988, 96 pages, softbound, $14.95
- **America's Pictorial Quilts,** Caron L. Mosey, #1662: AQS, 1985, 112 pages, hardbound, $19.95
- **Applique Designs: My Mother Taught Me to Sew,** Faye Anderson, #2121: AQS, 1990, 80 pages, softbound, $12.95
- **Arkansas Quilts: Arkansas Warmth,** Arkansas Quilter's Guild, Inc., #1908: AQS, 1987, 144 pages, hardbound, $24.95
- **The Art of Hand Applique,** Laura Lee Fritz, #2122: AQS, 1990, 80 pages, softbound, $14.95
- **...Ask Helen More About Quilting Designs,** Helen Squire, #2099: AQS, 1990, 54 pages, 17 x 11, spiral-bound, $14.95
- **Award-Winning Quilts & Their Makers: Vol. I, The Best of AQS Shows – 1985-1987,** edited by Victoria Faoro, #2207: AQS, 1991, 232 pages, softbound, $19.95
- **Award-Winning Quilts & Their Makers: Vol. II, The Best of AQS Shows – 1988-1989,** edited by Victoria Faoro, #2354: AQS, 1992, 176 pages, softbound, $19.95
- **Classic Basket Quilts,** Elizabeth Porter & Marianne Fons, #2208: AQS, 1991, 128 pages, softbound, $16.95
- **A Collection of Favorite Quilts,** Judy Florence, #2119 AQS, 1990, 136 pages, softbound, $18.95
- **Dear Helen, Can You Tell Me?...all about quilting designs,** Helen Squire, #1820: AQS, 1987, 56 pages, 17 x 11, spiral-bound, $12.95
- **Dyeing & Overdyeing of Cotton Fabrics,** Judy Mercer Tescher, #2030: AQS, 1990, 54 pages, softbound, $9.95
- **Flavor Quilts for Kids to Make: Complete Instructions for Teaching Children to Dye, Decorate & Sew Quilts,** Jennifer Amor, #2356: AQS, 1991, 120 pages., softbound, $12.95
- **From Basics to Binding: A Complete Guide to Making Quilts,** Karen Kay Buckley, #2381: AQS, 1992, 160 pages, softbound, $16.95
- **Fun & Fancy Machine Quiltmaking,** Lois Smith, #1982: AQS, 1989, 144 pages, softbound, $19.95
- **Gallery of American Quilts: 1849-1988,** #1938: AQS, 1988, 128 pages, softbound, $19.95
- **Gallery of American Quilts 1860-1989: Book II,** #2129: AQS, 1990, 128 pages, softbound, $19.95
- **The Grand Finale: A Quilter's Guide to Finishing Projects,** Linda Denner, #1924: AQS, 1988, 96 pages, softbound, $14.95
- **Heirloom Miniatures,** Tina M. Gravatt, #2097: AQS, 1990, 64 pages, softbound, $9.95
- **Home Study Course in Quiltmaking,** Jeannie M. Spears, #2031: AQS, 1990, 240 pages, softbound, $19.95
- **Infinite Stars,** Gayle Bong, #2283: AQS, 1992, 72 pages, softbound, $12.95
- **The Ins and Outs: Perfecting the Quilting Stitch,** Patricia J. Morris, #2120: AQS, 1990, 96 pages, softbound, $9.95
- **Irish Chain Quilts: A Workbook of Irish Chains & Related Patterns,** Joyce B. Peaden, #1906: AQS, 1988, 96 pages, softbound, $14.95
- **The Log Cabin Returns to Kentucky: Quilts from the Pilgrim/Roy Collection,** Gerald Roy and Paul Pilgrim, #3329: AQS, 1992, 36 pages, 9 x 7, softbound, $12.95
- **Marbling Fabrics for Quilts: A Guide for Learning & Teaching,** Kathy Fawcett & Carol Shoaf, #2206: AQS, 1991, 72 pages, softbound, $12.95
- **Missouri Heritage Quilts,** Bettina Havig, #1718: AQS, 1986, 104 pages, softbound, $14.95
- **Nancy Crow: Quilts and Influences,** Nancy Crow, #1981: AQS, 1990, 256 pages, hardcover, $29.95
- **No Dragons on My Quilt,** Jean Ray Laury with Ritva Laury & Lizabeth Laury, #2153: AQS, 1990, 52 pages, hardcover, $12.95
- **Oklahoma Heritage Quilts,** Oklahoma Quilt Heritage Project #2032: AQS, 1990, 144 pages, softbound, $19.95
- **Quiltmaker's Guide: Basics & Beyond,** Carol Doak, #2284: AQS, 1992, 208 pages, softbound $19.95
- **QUILTS: The Permanent Collection – MAQS,** #2257: AQS, 1991, 100 pages, 10 x 6½, softbound, $9.95
- **Scarlet Ribbons: American Indian Technique for Today's Quilters,** Helen Kelley, #1819: AQS, 1987, 104 pages, softbound, $15.95
- **Sensational Scrap Quilts,** Dara Duffy Williamson, #2357: AQS, 1992, 152 pages, softbound, $24.95
- **Sets & Borders,** Gwen Marston & Joe Cunningham, #1821: AQS, 1987, 104 pages, softbound, $14.95
- **Somewhere in Between: Quilts and Quilters of Illinois,** Rita Barrow Barber, #1790: AQS, 1986, 78 pages, softbound, $14.95
- **Stenciled Quilts for Christmas,** Marie Monteith Sturmer, #2098: AQS, 1990, 104 pages, softbound, $14.95
- **Texas Quilts – Texas Treasures,** Texas Heritage Quilt Society, #1760: AQS, 1986, 160 pages, hardbound, $24.95
- **A Treasury of Quilting Designs,** Linda Goodmon Emery, #2029: AQS, 1990, 80 pages, 14 x 11, spiral-bound, $14.95
- **Wonderful Wearables: A Celebration of Creative Clothing,** Virginia Avery, #2286: AQS, 1991, 168 pages, softbound, $24.95

These books can be found in local bookstores and quilt shops. If you are unable to locate a title in your area, you can order by mail from AQS, P.O. Box 3290, Paducah, KY 42002-3290. Please add $1 for the first book and 40¢ for each additional one to cover postage and handling.